AF412546

LIFT UP YOUR FACE!

LIFT UP YOUR FACE!

*What YOU Can Do, Throughout the Years,
About Your Face*

"Every man over forty is responsible for his face."

A. Lincoln

By

HELEN PEABODY

Illustrated by

Frederick Gibbs

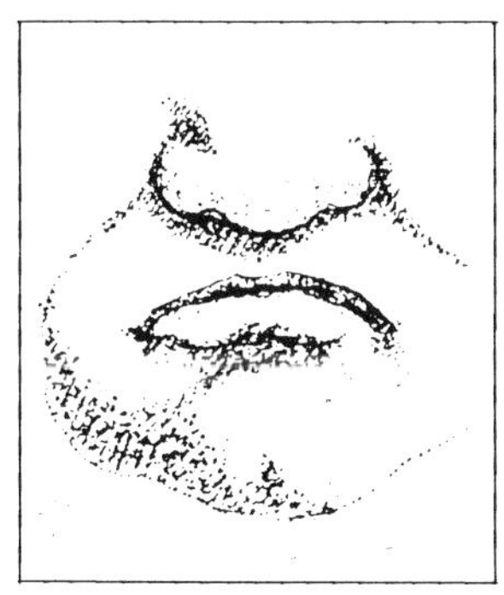 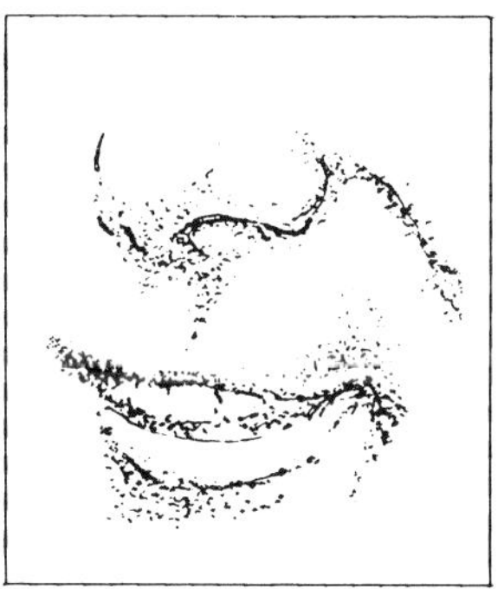

THE CHRISTOPHER PUBLISHING HOUSE
NORTH QUINCY, MASSACHUSETTS 02171

The Sequel to *Lift Up Your Face!* (Now in preparation)

Seeking to know ever more deeply the source of the face of beauty, Helen Peabody turned to studies of the philosophy of the East and fresh approaches to the Christian faith. Her book took a new turn, developing a partner volume called *The Sequel to Lift Up Your Face!* There, the flesh and blood face becomes the lamp of the spirit; there, we get the inside story of what we can do about our faces.

Further information about this forthcoming second volume can be obtained by writing to the author in care of this publisher.

A Challenge to Old Age

One of Abraham Lincoln's advisors urgently recommended a certain man for a post in the President's cabinet. When Lincoln declined to follow the suggestion, he was asked to give his reason.

"I don't like the man's face," the President explained.

"But the poor man is not responsible for his face," his advocate insisted.

"Every man over forty is responsible for his face," Lincoln replied.

Frances Parkinson Keyes
in "This Week Magazine"

This book is written
to men and women equally;
to all ages;
to people with all shades of skin:
it is written to us all,
for the wish to look our best
all through the years
is a universal one—
part of being alive.

AN APPRECIATION

This book has been coming, on and off, for a period of thirty years. As I look back over the many friends who have helped in its preparation, the following stand out in glowing colors, all sixty-eight of them:

—Kathryn Burbank, Mrs. Monroe Burbank, costume designer, who finding this manuscript in a bottom drawer, believed in it, worked on it, and set it going again;

—John Gallishaw, writer and author's critic, whose certainty that this book would be translated and go widely around, and his creative criticism, propelled me on my way in no uncertain fashion.

—Helen Day, Mrs. John Franklyn Day, for her sustaining faith in this manuscript and in the forthcoming sequel which contains the deeper story;

—my cousin Cornelia Runyon, Mrs. Charles Montgomery Runyon, sculptor, for her refreshing enthusiasm for this text throughout the years;

—Emily Lawrence, Mrs. Theodore Newton, actress (La Belle Emilie of these pages) for her never flagging interest and critical help;

—my cousin Richard A. May, business executive, for his sustained concern and practical help in the selling;

—Robin Prising, writer, publisher, understanding critic, who has been wholly successful in setting to work in his own face the ideas of these pages.

—and all the many others each of whom has contributed some vital element in helping this manuscript along from its rough beginnings to its present form:

Ethel Amory, Winnifred Barrett, James S. Best, Lise Bieberman, Mary Connell Bray, Lucille Buhl, Margaret Bullowa, Barbara Burr, Helen W. Congdon, Isabel Conklin, Vivian Cooper, Margaret Darst Corbett, Julia Cunningham, Mildred Gignoux Downes, Virginia Ehrlich, Abigail Adams Eliot, Frederick May Eliot, Martha May Eliot,

Helen E. Ellis, Amelia F. Emerson, Eugenia Everett, Charles Fager, Guido Ferrando, Dorothy Canfield Fisher, Mildred Buchanan Flagg, Margaret Grandgeant Gifford, Gertrude W. Hoffmann, Mary Peabody Hotson, Roberta Hughes, Robert A. Johnson, Jessie M. Jones, Pauline Kramer, "Maggie" Letvin, Isidore and Bessie Levitt, Grace and Fay Luder, Mary Lou Maisel, Suzanne Miller, Claire Murray, Frances Taber Neff, John A. Nimick, Jeanne Marie Pearse, Rosamond Folger Pierce, Susan Regnery Plotnick, Patricia Read, Helen Robinson, Eloise Cummings Simpson, Abigail Eliot Smith, Robertson Smith, Raymond Spencer, Polly Starr, Rosalie Stewart, Renée Thornton, Tish Thomas, J. Donnell Tilghman, Margaret Tims, Janet Turpin, Nora Unwin, Gertrude Wagner, Jean Yeomans, Julia Day Yeomans, Ludmila Zielinski, and so many others. . . .

To each one, my profound appreciation, and loving greeting on the eve of publication.

Every one of us is destined to wake up some morning and, catching a glimpse of himself in the mirror, hear his heart cry out, "Dear God, how old I look!"

Sometimes it seems as though we could accept all other symptoms of growing old better than *growing old looking*.

The importance of the individual's facial appearance is not in question. That is an established fact the world over. But in spite of our concern over the face, the flesh of it is the most neglected, misunderstood part of the entire body. Why is this?

How many people are there, do you think, who know that the way the flesh stands on the facial bones is subject to their own control? Very few, I am sure you will agree. But it is controllable, to an extent undreamed of in the past, and this book is written that you may have proof.

There is a lovely unexpected dividend here: since the face is the symbol of the whole of us (central switchboard of the sensory organs, and outer expression of mind, body, and spirit) the help we manage to provide this extraordinary area will speedily reach through to the intangible side of us, and give a lift to the total personality.

Before we start off on our voyage of discovery let me tell you a story. This episode comes from a time when I was employed as head of the facial department in one of New York's fashionable beauty houses; it is a true story and about a woman, but in a different setting it could be equally said about a man. What is more, it is about a white woman: but color obviously is a nonessential. The story would be equally valid for a human being of any racial shade.

What of My Face?

When the curtain first rose upon this dramatic essay, one lighted area emerged from the shadows—a dressing table with a pair of crystal lamps and a mirror above. From the depths of this mirror the face of a woman was searchingly reflected. Here was proof in plenty that all we need for a whole drama is one living human face.

The audience for this drama was two people, the owner of the face—now seated at the dressing table awaiting her preliminary analysis, and myself, looking over her shoulder into the glass. It would have been difficult, I believe, to have determined which of us was the more absorbed. Surely this face was telling both of us of the wreck of a beauty. But let me begin by telling you about the hair.

The lady attached to the hair had taken the pins out and the red-gold flood of it was making its way over her shoulders into the darkness. It was without question the most beautiful hair I ever hope to see. But such hair belongs to the stage! I was startled to think of its wearer carrying around all the weight of it on her head every day of her life—especially at a time when other women were enjoying the lightness and freedom of short hair. . . . Obviously, the hair was the one aspect of this woman's beauty that had not succeeded in eluding her.

The face, the skin, though fine textured, and without a blemish, sagged everywhere that the flesh of a face can sag. In so doing it presented a linear pattern sufficiently pronounced to outweigh the structural pattern beneath. The mouth drooped and was embittered. The eyes that sought for mine were without illusion. As though in a trance she of the face placed her fingers at key points of temple and angle of jaw and drawing up the loose flesh said to me, "See what it does for me when I hold up my face!"

She was considering having her face lifted. She was ready to go to one of the leading surgeons in the field and pay his price (in those days very high indeed). We talked it over as I gave her a treatment and in the sympathetic atmosphere of the studio she told me something of the story of her life. Even without the story, my fingers on her face told me how long it was since she had known constructive living.

But there was something about this woman that I very much liked. And I was strongly under the impression that if she had known twenty years earlier what I then knew about faces, she would never have allowed her face to go to pieces like this.

She never did resort to the surgeon. She stayed with me and persevered in the ways that I suggested. When I last saw her she looked like an entirely different person—a woman of fifty still, but a much younger fifty, and what is so much more important, she was once more beautiful to look at. The tensions in the forehead had vanished. The sagged-down effect of the flesh, like clothes hanging on a line, no longer predominated, leaving the observer once more free to perceive the delicate beauty of the proportions. The eyes were once more really open, revealing their wonderful chestnut color.

She had cut the mass of her hair. At the time she left New York she was wearing it dressed high—a few soft curls on the top of her head. In this position the color related itself to her face even better. *She* at last was wearing *it*, and for any beauty accessory that is the right way around. Wherever she is now I know that many an artist would wish to paint the face that looks out of her mirror. I trust that she will some day enjoy this book, for it was unquestionably her interest, her touching receptivity and finally her progress and her delight in it, that encouraged me to develop this thesis into a working whole.

So let us take the plunge now and look immediately into the situation for ourselves. In the first chapter, *You Are The Sculptor,* as throughout the whole book, I shall reach into a variety of fields to gather my material: art, with its portraiture, design, make-up for stage; anatomy, physical culture; psychology, philosophy—and any other branch of human thinking or experience that can help us realize our muscles as modelling clay and ourselves as sculptors, as artists. At the end I shall roll all my findings into one condensed little what-to-do as the years go by, not only to preserve a good appearance, but significantly to enhance it.

You are about to discover that time and deterioration do not necessarily go together. On the contrary, the years can turn out to be an opportunity to develop, each in his own body workshop, the finest bit of pattern in the visual world—a face that is firm, "up," expressive and exquisite in age.

So let us give ourselves up to this project for a while and in this rushing world relax, take our time, and enjoy ourselves.

THE LIVING HUMAN FACE – A VISUAL MIRACLE

Portrait of the poet-philosopher Rabindranath Tagore. Photo by Bettini of Rome and New York.

Chapter 1

YOU ARE THE SCULPTOR

This book is about face muscles—a territory still largely unexplored. The surgeons trim these muscles in an effort to keep them young looking; but how many understand them?*

The muscles of the face are the costume of the face bones; the way we wear them tells the observer who we really are. But let me begin this study with a consideration of the human face as a work of art.

Every face is like a little architectural edifice of some sort—a little one-man temple. The central column is the nose; the arches are the brows; a base for the column of the nose is provided by the lines and forms of the mouth; the substructure is the chin. Column and arches together support the roof or the dome above. Here we have a bisymmetrical pattern that is quiet, solid, suggestive of enduringness and peace.

The variety in basic facial styles is just as demonstrable as the variety in architectural styles: the Classical, the Gothic, the Rococo, the Modern. If in addition to such basic considerations of design we add the incandescence, the dance of expressiveness of the living face, we just begin to understand what a visual miracle any face turns out to be.

In a portrait of Tagore in his later years, we can easily identify the little one-man temple. The fine eyes look out from beneath the arches with a timeless serenity. In such examples we realize that faces beautiful in age make up the finest bits of pattern upon which the human eye can rest—masterpieces of nature and spirit in collaboration.

---

*I have no need to be at odds with the face surgeons. On the contrary, my hope is that this book will prove useful to them in their work, too.

The Expressing Face Muscles

The muscles of the face are strategically located—like sentinels at the gate between the inner and the outer man. We might say that they were like the glass at a window, the thoughts and feelings inside the house look out through them; and then again, those of us on the outside can look through the muscles of the face of another and see the thoughts and feelings inside.

Pause for a moment to think what it would be like for us if our faces were as relatively immobile as the rest of our flesh—living masks, it is not only the other fellow who would be out of luck, for by our faces he makes excellent guesses at our feelings, but we ourselves would feel almost as handicapped as though we were deprived of one of our five senses. Surely it must have been our need for these diversified little muscles that brought them into being. Out of all living creatures only man is so lavishly supplied with muscles of expression. By means of them he presses out the finest shades of feeling of his complex human heart.

This brings us to the necessity of looking into the emotions activating these muscles. As we know, feelings do not come from our obvious, conscious experience only; many of them, by far the greater part, come from the deeper layers of mind and heart. And the problem turns out to be: does the expression of all these feelings, surface and hidden, over the years, give the face muscles the balanced workout they need to keep them in good condition?

The first step is to find out what actually happens to the face muscles under the pull of the emotions. The happenings here are so revealing of so many points of interest to the beauty seeker that we need to consider them with the utmost care.

Of course everybody, more or less, knows about muscles; the important thing for us is to have a common approach to the matter. So I shall lay before you now a short study of muscles, all of them, body and face; then we will focus our attention upon the less well known muscles of expression.

All Muscles as Functional Tools

Let us begin with one of the more familiar body muscles, the one on the front of the upper arm. Everybody knows that one.

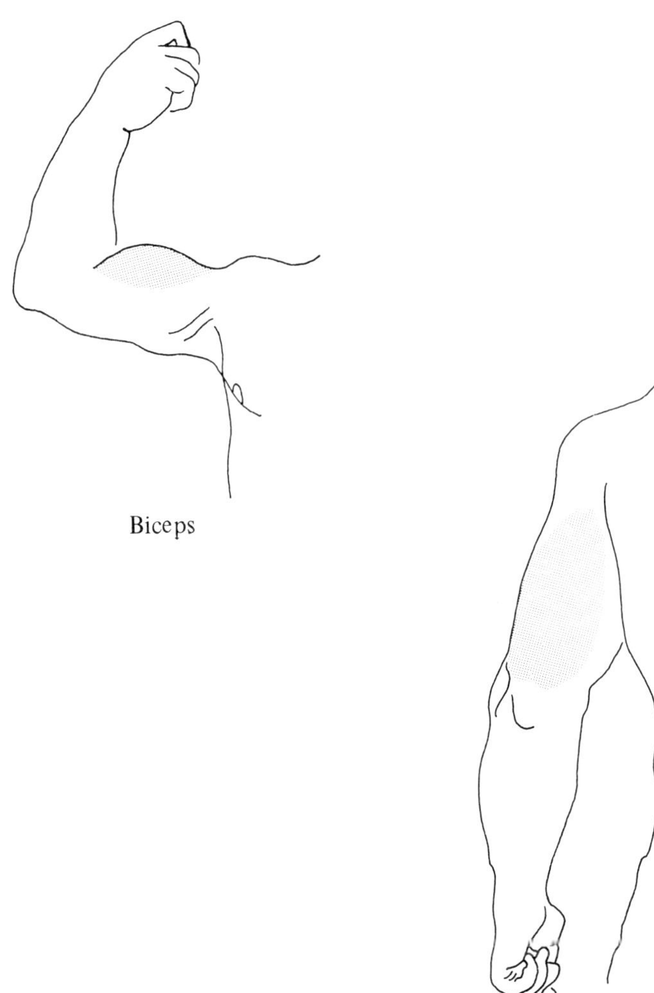

To achieve motion all muscles function in pairs of opposites, like the biceps and triceps of the arm: the biceps pulls in one direction, the triceps pulls back again; pull, pull—like a couple of men at work on a two-handed saw. No muscle pushes!

The Great Biceps

The lower end of the biceps is fastened to the head of the big forearm bone, just below the elbow joint. By reason of the fact that this muscle crosses over the joint, it has leverage upon the forearm. Place your right hand over your left biceps as though you were getting ready to show off how strong you were, then tense up the biceps. Obediently, up comes the forearm. Relax the biceps. These simple movements show us that muscles achieve the special motion they are designed to accomplish by means of *tensing up.* Our muscles are strictly pull mechanisms; they never push us around. Even the tongue gets thrust out by muscles operating like pulleys.

So, tensing up is a primary function of all muscles; however, it is not the only function. All muscles have a dual function: to tense *and to relax.* This important dual function of muscles may well be described as *a pair of opposites.*

Everyone knows that we human beings see our world in terms of pairs of opposites; we understand that we have, in us and about us, a dual world. We find such pairs as good and evil, north and south, happiness and sorrow, male and female, disease and health, youth and age, tears and laughter, on and on. . . . The pairs of opposites provide us with our stability and balance. We can hardly imagine our world without them; but they are not all, there is a value between and above them as you will see.

As muscles tense up they grow shorter and thicker; they plump up, as it were. *As muscles relax they elongate.*

Then again, muscles that stay tensed too long develop a condition known as *supertension,* with the familiar pinching up; and muscles that stay relaxed too long develop a condition known as *over-relaxation,* with the familiar flabbiness.

Two Men at Work on a Two-Handed Saw.

So now we have these two important pairs of opposites: the normal pair that represents right functioning, and the abnormal pair that represents excess:

> tension and relaxation,
> supertension and overrelaxation.

We will make use of both of these pairs, often.

The splendid concluding point here is that perfect balance between tension and relaxation in the muscles produces the cherished *tone*. This tone is a natural characteristic of youthful muscles. However, a by-product of this study is the discovery that perfect tone is not, necessarily, a characteristic of youth alone.

The Partner Muscles

Once again, tense up the biceps. Up comes the forearm. The question now is: how to get the forearm back to its original position. Discounting the pull of gravity, just relaxing the biceps does not produce any movement in the forearm. We have to look to another muscle, similarly disposed, on the *back* of the upper arm, also working by contraction, to draw the forearm back to its original position.

> From this action we see that it takes two muscles operating on opposite sides of a joint to give the joint its full quota of motion.

One muscle, or set of muscles, pulls the piece of us in question forward; another muscle, or set of muscles, pulls it back. It is like a couple of men working on a two-handed saw: one man pulls forward, the other man pulls back; pull, pull.

Muscle Adaptability

Another characteristic of all muscles in general is their unfailing *sensitivity*. Sensitivity to stimulation is really a characteristic of being alive, and so, of course, it applies to muscles, too. Muscles adapt themselves to what comes, exactly as plants do—good, bad, or indifferent. It is called muscle adaptability. In this responsiveness lies our hope.

Who Is Boss of the Muscles?

This is a question that must be answered in connection with each of the muscles that make up this study.

All body muscles are classified in one of three groups: (1) First is the crew that works over the bones. Most of these are long straplike affairs, that are in charge of bodily locomotion; they are controlled by our conscious faculties and are called *voluntary muscles*. The biceps, obviously, is a voluntary muscle. (2) Then we have the crew that is in charge of the action of the organs—the beating heart, the breathing lungs; these are controlled by nonconscious faculties and are called automatic or *involuntary muscles*. (3) Lastly we have the crew in charge of expression—the face muscles; these are controlled by a bewildering combination of voluntary and involuntary faculties.

The Muscles of Expression

These muscles are not only just as sensitive as other body muscles but, because of the complex system of nerves in the face, are much more so. We must not forget that these muscles, equally with those of the body, operate exclusively by contraction.

The first point of difference between face and body muscles is that the face has only one workable bone joint, the hinge of the jaw. Yet we find the muscles of the face also functioning like a two-handed saw. Here the action takes place over a relatively flat plane; but just the same we see one muscle, or set of muscles pulling in one direction, and another muscle, or set of muscles, pulling back again. The subtlest expressing job a face can achieve is produced by muscles in pairs, working opposite each other.

The Feeling Climate for Face Muscles

Our bodies, even though we have no time or even inclination for deliberate physical exercise, just in the ordinary round of life, have considerable opportunity for action. But all the time this physical activity is going on, what is happening to the face muscles, do they get any good out of it? Only indirectly, surely, for there they are sitting up on the face bones being chauffeured all over the place. It is true that the face muscles are no help in locomotion, washing windows, or playing the piano; but, as we know, they are by no means inactive.

Up there in the face they carry on their own endless dramatic activity. Ordinarily this activity does not produce a balanced development for the face.

1. *The Worry Man and Supertension*

Let us watch a certain man who is just emerging from his office building onto the city street. His lips are compressed, the corners of his mouth turn down. There is a deep furrow in his forehead, his eyes are narrowed. He doesn't need to look out, he knows this street so well he could walk in it blindfolded.

This man is an able person; but just now he can't figure life out and he is heartsick that he finds himself unable to make a better living. Let us call him the Worry Man, and keep his face in the reference file of our imagination.

If in this fellow's face, at the moment we have chosen to study it, many muscles are being held in a state of supertension, the rest of them, the ones designed to pull back the other way, are in a state of inactivity—as complete an inactivity as it is possible for muscles to have and their wearer still be alive. Inactive muscles do not show up immediately with obvious results as do tensed ones; but over the years they steal forth in jowls, curtains over eyes—all the aging face draperies.

2. *The Gloomy Man and Overrelaxation*

We will now look into another man's face. I passed him on the street the other day. I wanted to take him by the hand and present him to my stage makeup class as an example of the man who never smiles. How pitiful that there should be such a face anywhere. Everything in it that could droop was drooping; he looked like a good old bloodhound. But faces of hounds, like willow trees, are beautiful because drooping is their rightful pattern. Let us keep this man's face, too, in our reference file. We shall call him the Gloomy Man.

I want you now to try on these two expressions, worry and gloom, so that you can refresh your memory as to the feel of them in your face. Let us begin with the Worry Man. Imagine that you are standing in the wings of a theater getting your mind and your face ready to go on stage and play a scene of intense anxiety. In response to the worry signal travelling over the nerves from the brain a whole set of muscles tenses up, and contracts the face inward toward the nose and also down. You readily get the feel of it. In the expression of worry one set of muscles is making an immense effort.

And now let us imagine that you are alternating in the role of the Gloomy Man. We will take this man's emotion in its extreme form—despair. In this case the mind and heart, with its face, have given up. There is no action whatsoever in the face muscles; you let go of them, you just let them hang down. An actor beyond you, gravity, takes over and pulls them down.

In these two expressions, worry and despair, we experience, directly, our second pair of opposites, supertension and overrelaxation.

So we find that worry and all degrees of fear—long continued—produce supertension in one set of the face muscles, with the overall tendency to pinch the face inward toward the nose; and, conversely, despair, gloom, apathy, and all degrees of negation—long continued—produce overrelaxation in all of the face muscles, with the overall tendency for the face to sag. Most older faces show some of the pinching in and some of the sagging down, both.

But, needless to say, fear and all that company; despair, negation and all of these, do not constitute the sum of human emotions. There is a third feeling group, the joy group and all of its collaborators—happiness, peace, confidence, merriment, laughter, hope, love, joy.

3. The Face of Joy and Perfect Muscular Tone

I am recalling the face of a certain young mother holding in her arms her newborn son. . . .

I want you to try on this expression, too. Imagine that you are already on stage, and that the moment finds you, mother or father, extending your arms to receive your newborn child.

If you are a good actor this is what is happening in your face. All the muscles that were lying idle during the expressions of worry or gloom, fear or despair, now rise up and tug like anything. This time, Oh glorious magic, the face opens like a flower!

Yes, *the joy muscles lift the face, widen it, and by so doing subject it to a molding and shaping in the direction of the ideal form.*

Here now is a point I want you to consider with great care: at the time when the joy muscles are in action the worry muscles are receiving their stretch-back-in-the-opposite-direction, their beauty treatment—which prevents them from falling into posture creases!

The experiences of happiness tend to produce in the flesh of the face the perfect muscular *tone,* the nature-beauty we spoke of earlier that comes from balance between tension and relaxation, but this is not all:

When the experience is one of joy, all or in part a spiritual thing—which it always is, otherwise it is not joy—there comes to the flesh of the face, also, a highly sensitized, illuminated quality such as we find in the face of Tagore, or better still, once in a while in the faces of those about us. This lovely quality of the flesh, expressed in terms of our pair of opposites, tension and relaxation, could be described as *the golden mean raised up between them.** But whatever words you may use to describe the structure of beauty, you can count upon it that any person who has beauty in later years has had a more than passing acquaintance with joy in the heart.

And Now to Summarize:

> As they stand, these three expressions: worry, gloom, and joy, represent the three extremes, of the effects that the emotions have upon the flesh of the face.

We know, of course, that a person caught up in life's feeling currents is not concerned in the slightest degree with the appearance of his face. And that is right. I am no advocate of self-consciousness.

---

*This philosophical concept of the golden mean that is raised up between a given duality or pair of opposites, comes to us from the Orient—the Tao—that flourished in China thousands of years ago and is still with us to this day. (*The Way of Life According To Laotzu* by Witter Bynner, ed., New York: John Day Publishing Co., 1944.)

JOY LIFTS MUSCLES — ALL OF THEM

Photo by Dorchester AVCO, courtesy of Laymens' League, Unitarian Universalist.

When the current is on is no time to repair the lamp. What I propose to do is to clear the way for the facial muscles so that they may function freely and expressively in the triad where they feel most at home—tension, relaxation, and balanced tone. To leave our face muscles at the mercy of the emotions turns out to be very unrewarding in exactly the same sense as it would be for body posture. We need first aid for faces, and a first aid that will also stand by us through the years.

One day, as I stood on a station platform waiting for a train, I noticed a woman standing a few yards away, her back towards me. I admired her trim figure, the way she held her body and wore her clothes. Then she turned toward me, and seeing her face, I was shocked to find that she looked twenty years older than I had expected her to. She had done a fine job with her body, but her face had eluded her.

Good Face Posture

Our working hypothesis then is as follows: *The face appears at its best in the expression of joy.*

From here we set out in the following pages to clarify what the face muscles are doing in the expression of joy, so that any one of us can make use of the structure of it more or less continuously even in depressed or anxious times. Good face posture is to the face what good body posture is to the body; nothing less, and considerably more, as you shall see.

Good face posture unsupported by any immediate emotion from the joy group results in an expression of friendliness and alertness that any one of us would wish to include under the heading of good grooming. It is the lifted face (see definition below). Holding the face in good posture will result in a face ready to smile—you don't have to take up a yard of slack before you can get going—the sub-structure, muscularly speaking, of the vivid smile. A face unhampered by bad posture habits becomes an uncommonly sensitive instrument of expression. Here you will find the actors who can play many different kinds of roles. You will discover for yourself soon enough the influence of these matters upon the personality.

The definition of Good Face Posture, then, goes like this:

> *Conscious* use of the face muscles as we find them function-
> ing, *spontaneously*, in the expression of joy.

We have already seen how faces look in conditions of muscular functioning that do not contribute to beauty. The Worry Man was our example of overworked muscles; the Gloomy Man of underactive muscles. Our search is for the perfectly balanced tone, the golden mean between these two. Later on when we have found out just what this muscular functioning is for the whole face, it will then remain for us

> to get the habit of it, until holding the face muscles right,
> again becomes a *subconscious* process.

At that time we will need to give the matter very little further attention. Doing the right thing for the muscles will then have come to be as easy and natural for us as doing the wrong one was formerly. We are going to have the wonderful fun of finding out that nature and time will both play the game on our side, the side of beauty, when we understand their laws.

In the chapters ahead I shall take up each feature separately, pointing out what is going on in most of our faces as we grow older. Where there is supertension I will show how to relax these muscles; when there is overrelaxation I will show how to reactivate them. It is for me to clarify further that both processes are going on in our complex faces at the same time. I would like, at this time, to pass along some remarks made by a distinguished New York surgeon as he completed a reading of this book still in manuscript form.

"What you say about muscle adaptability," he said, "is something we surgeons have known for some time, and have been putting into practice in relation to the muscle of the body; but it is extremely interesting that we did not think of making the application to the ordinary processes of the aging face.

"Take for instance the round-shouldered boy," he said, "I have seen these back muscles in my surgery, ever so much longer than normal. You can see how these muscles no longer serve to hold the boy up; they have accommodated themselves to his posture and serve only to hold the bent over posture.

"Or take the reverse of this," he went on. "Take the girl who all her life has worn high heels. The muscle at the back of the leg becomes shortened so that she can no longer be comfortable barefoot. Nature figures out that she does not need the full length of the muscle; scar tissue forms and the muscle starts all over again, on the basis of the reduced length. Any muscle subjected to supertension shortens eventually."

There is, of course, a parallel between body and face muscle adaptations. Like the boy's overlong back muscles, certain cheek muscles, having been insufficiently used, elongate and droop over the jaw bone, hence the jowls. Conversely, like the shortened muscle at the back of the girl's leg, the small muscles in the forehead that pull the eyebrows together grow constantly shorter under supertension, hence the scowl wrinkles.

These muscular doings are examples, both in body and face, of muscle adaptability that is producing wrong results for both health and beauty. As such they are significant in helping us figure out the harmful tendencies; but the positive side of muscular adaptability provides us with the heartwarming constructive side of our task. It may be summed up in the one familiar line: as you hold it so it grows.

Chapter 2

PATTERNS OF OLD AGE

And now before we settle down to our study of individual face muscles I have a surprise for you—a little television show.

This production has but two featured players, a makeup artist for stage and a young model. The make-up artist will paint upon the model's face the whole development of the aging process, discussing it along the way. Mankind has for so long taken this appearance for granted I figured a candid review of the situation would help wipe away some of the prejudice and open the door for a fresh experience.

A young actress friend of mine will play the model and I shall play the makeup artist, a familiar role for me. I believe the men and women in the audience will, equally, enjoy watching Emily, lovely as she is, standing in for both sexes in this encounter with the "years." "Age" comes to all of us in much the same way.

So here we are before you now on the technicolor screen. Let me introduce you to La Belle Emilie. I am sure you will agree with me that she has a perfect face for our requirements—a full oval. Her eyes are blue as gentians; her well-proportioned nose turns up ever so slightly, contributing to the gay lift of her face. Her hair is the color of a copper moon. She will sit here at this dressing table where I have lights and makeup all prepared.

And now I shall roll Emily's colorful hair up and out of the way in a becoming chignon, tuck a towel over her shoulders, and after a light touch of foundation paint over all, starting right in at the eyebrow. What kind of an eyebrow is this one of Emily's? It suggests the Gothic surely. It shoots out from the nose, rising in sprightly fashion to a point a little past its center, then drops away in a lovely curve. It is a dynamic line imparting life and animation to the face.

Eyebrows play an important role in the facial design; they are also important in characterization, so what happens to them as the face grows older is worth considering. Here it is: slowly but surely, the eyebrows drop down.

Let us examine some by-products of drooping eyebrows. People who start with straight eyebrows and not much space between brow and lid sometimes end up with the eyebrows appearing almost to fall into their eyes. Then too, character traits show up when eyebrows, still young, start developing a diagonal droop. If an actor in a certain role is supposed to represent an unhappy character, I change his eyebrows so that they have this wistful slope. A theater full of people, though they will not be aware of the reason, will get the feeling of it.

Let us now focus our attention upon the space between the eyebrow and the upper lid. This little area where apparently nothing is going on, does for the eye the same practical service that a beautiful white mount does for a water color: it provides the area of enhancement between picture and frame. And what happens to this little space in the aging face? The eyebrow, since it is on the way down, crowds this small but important design element right out of the picture, with the result that the eye is diminished in luster.

Now, Emily, I am going to block out your eyebrows entirely and, skipping forty or more years, give you a pair of new ones such as you might have at sixty-five. Makeup artists for stage stick down the natural brow with moustache wax and paint over it; but it is difficult to hide the natural brow from the camera's all-seeing eye. I shall do this sticking down one side at a time so that we may compare.

There, now that this eyebrow has disappeared, you can better appreciate its value in the facial pattern. Without the eyebrow to hold and frame it, the eye looks as though it might fly out in any direction. And the nose, cut off from its relationship to the design by means of the brow, a pillar without an arch, appears to be stranded like an old column in a meadow after the exodus of the World's Fair.

Here comes now the "older" eyebrow, a little straighter, and down nearer the eye. Isn't it surprising what that does for Emily, or to her? The design has changed so little, and yet the whole aspect of her face is different. The eyebrow, less spirited, more conservative, seems to have taken the joy right out of this side of her face.

Before we go on, let us see if we can find out where the lost space between eyebrow and upper lid has gone. It can't have just vanished into thin air. Yes, here it is. Do you see this little curtain that is coming down from above, that partially conceals the upper lid and veils the outer corner of the eye completely? By tricks of light and shade it is possible to simulate this small curtain.

Please do not think that as we go along I am forecasting Emily's future. Not at all—I am outlining on her face a trend, that, assuming individual differences, is taken for granted in the older faces of our day. We may expect that, due to the large component of joy in her personality, Emily herself will be free of much of it.

We are not yet through with this eye, however: just here below it in the structure of the lower lid, I will paint in the "tired circles" that cradle puffy or light places.

That done, now let us compare the "aged" eye with the natural one on the other side. Isn't it extraordinary how much smaller the "old" eye looks and how much less bright? Yet its actual size and brilliance has, of course, not been tampered with. Both these effects indicate the extent to which visual appearance is an optical illusion.

Now what about the familiar "old age lines?" Let us look first at the smile creases radiating from the outer corners of the eyes, the "crow's feet." Give us an exaggerated smile, please, Emily, so I can find your own. And now push up the forehead to show the horizontal wrinkles there. These are easy. I just indicate them lightly.

And now directly between the eyes we have the "scowl wrinkles." You see Emily has two strong vertical lines here when she calls for them. They seem to be just lying in wait for her. Actually we can call up almost every one of the aging face symptoms with a specific expression. But that's enough for the lines just now; right here under the brush, demanding attention, comes the nose.

What happens to the nose in the aging face? The story would seem to be about as follows: whatever the nose is at twenty, it is likely to be more so at sixty. Is it "fine?" It may get a little peaked. Is it Roman? It may turn into a beak. Is it round on the end? It may grow a little bulbous, I am afraid. Emily has one of those neat, in-between noses that is likely to change very little. I shall merely accent the sides with a touch of shadow.

We can proceed on down the face more broadly now. The flesh, less firm as the years go by, more and more reveals the contours of the skull bones beneath it. To get this effect I shall drop delicate shadows in the temples, under the cheek bones, and about the muscles of chin and throat, then return and pick out the prominences with highlights.

Having prepared the cheeks, we are now ready for the nose-to-mouth lines. I call these the "middle-aging lines." What! Emily, you don't like these elegant Mid-Victorian drapes? See—when she smiles they look like curtains of a miniature stage drawn up to await the prima donna's entrance.

The next step is to paint away the mouth's youthful fullness. Emily is going to like this even less. Since we are skipping so many years it will take delicate maneuvering to conceal the youthful fullness of her lips. But there they go, blocked out beneath flesh-colored grease paint.

And here comes the firm, more level line of the new upper lip, extending more to the side and drooping a little at the corners. That's too bad, isn't it? It almost seems as though this droop were lying in wait for her, too. See how easily it falls in here?

Now let us see what is happening to the line of the chin and jaw, the "contour" that in Emily's young face is as neat as the end of an egg. The sag of the flesh at either side of the chin makes the face look so much squarer than it used to. I assure you that there is nothing more characteristic of the feminine middle-aged face than this squaring off at the jawline. Stealthily, imperceptibly, as the muscles grow flabby, the cheeks steal down, and finally at their lower borders, a highlight supported by a shadow breaks across the clean line of the jaw, and (to the observer anyway) youth is gone.

From the jowls to the throat is but a carry on of the same process. I shall pick out for Emily the two major draperies of the aging throat just below the chin, and then the smooth column of the throat will be gone also. A little powder now to set the paint, and I shall shake a bit of talcum into the hair, and there she is: Dame Emily!

In this makeup Emily has both a delicacy and a dignity that are very lovely; but the look in her eyes reminds me that these first "age" makeups are hard to take. I once saw tears stealing down a young actress's face when she first beheld herself in such a role.

Taken altogether, some of the effects of this particular representation of an older face are caused by a pinching up, and some from a drooping down. It would seem that as the years go by the face takes on some of the aspects of a badly laundered woolen dress—it shrinks and sags in all the wrong places.

And that finishes this demonstration. You have been a wonderful model, Emily, and we thank you. Here is cold cream and cleansing tissue; now you may take off your mask. . . .

I personally believe that the aging process just described is the product of a lack of understanding of the marvellous structures of body, mind, heart, and spirit that we have been given to counteract them with. I believe that the time will come when this type of appearance will no longer be considered a necessary partner of the years. . . . Look at Emily's young face coming back again, will you! Isn't that refreshing, and with that smile, it is like the sun coming out on the daffodills after the rain.

Model without make-up   Grammar School   High School

Debutante   Young Mother   Older Mother

Granny   Great Granny   Just for a joke!

AGING SERIES: Make-Up by Helen Peabody
The model is Patricia (Coppal) Poindexter. Photography by Max Munn Autrey of Hollywood.

Chapter 3

MAINTAINING A JAUNTY SKULLCAP

The essence of beauty is like quicksilver; what we have in youth never stays captured; we have to keep winning it over and over again every second. If we hope to retain the clear forehead, the open eyes and the rightness of nose and cheeks we have to learn how to set the scalp-and-forehead muscle to work and work it hard for the rest of our lives. Compared to other muscles of expression this muscle is gigantic in size, in potential strength, and potential influence for beauty.

I figure that a human being could well start himself off as an artist by looking in the mirror for the first time with an impersonal eye. This face before him, he understands, or let us say he does, is his greatest art treasure. He can see, even now, that there are things he could do, that he should do, to bring this picture into more perfect harmony. As he thinks these thoughts he has begun to participate in creation—to be an artist.

I believe that every man contains within him a sense of design and a feeling for good proportion. How much of this is due to individual training and experience, and how much to creative intuition we need not settle here.

Let us begin with the manner in which the facial design produces its effects. In order to do so we shall avail ourselves of the classical standard of measurement, not because this scheme of things is ideal for any one of us—where would we be without the racial variations on the theme of beauty, not to mention the individual—but because classical symmetry and balance makes it well suited to be the standard, or point of departure.

The classical oval is no less a form than the good old-fashioned hen's egg. An egg is a remarkably interesting shape, and held big end up presents an area that could not be improved upon as a background for the human face. This perfect oval places emphasis where we like to see it, upon the intellect. It has the quietness and stability of a bisymmetrical form, with unity as its outstanding characteristic, and yet it is full of variety, changing at every moment throughout its entire length and breadth. Interestingly enough these remarks are such as could be made about many great works of art. We can ask for no greater proof of the excellence of the oval as a form than to see the way men—the ones blessed with good proportions—can "get away with" their bald heads.

Next we look into the layout of the features upon this ideal shape. Let us get out pencil and paper and draw a three-inch oval. Down the center of this oval we now draw a vertical line. Crossing this vertical at one third of its length, we then draw a horizontal line. These two lines, vertical and horizontal together, give us the pattern of a cross. In the design of the face, the nose provides the impetus for the up and down part, and the eyebrows (the brow) supply the emphasis for the horizontal. The spot where these two powerful lines intersect, at the top of nose and between eyebrows, is the dynamic central point of the pattern.

This type of cross set within the oval is the foundation of face design. From here on, as we bring in the features one by one, we shall see the facial pattern building up, back and forth and around, in fascinating rhythms. We can already see what a distinguished location this basic design offers for the eyes; no more expressive position for them could have been devised.

The Ideal, the Expressing Forehead

I once knew a man, an extremely intelligent person, with a fascinating, creative mind. This man was a writer and newspaperman. He had a pronounced receding forehead. This experience prevents me from saying that a lofty intellect is invariably housed in a lofty dome; but as we search the faces of our fellowmen we find that it does, indeed, often happen that way. See the portraits of Tagore, Robert Ingersoll, Sir Walter Scott, to think of only a few. However, I do say, unequivocally, that if any one of us were casting a play and trying to fill the role of a great musician, sculptor, scientist, whatever, we would pick an actor who had not only sufficient intelligence to be convincing in such a role but one who looked the part, and that would be one with a good dome above the ears.

The Cross Within the Oval —
Foundation of the Face Pattern

While we are discussing the ideal head, the dome of it, I mean, we must not fail to mention that this feature's beauty lies not only in the proportions of the skull bones but in the fact that it is ornamented with a lustrous crop of shining hair. And here comes my first opportunity to mention color—red, russet, black, honey-colored, brown, golden, white—how thrilling are the colors of the hair!

The dome's lower border, the entablature, is the forehead—one of our most lovely and impressive features. In design the forehead provides the necessary quiet area, or "mount," between the frame of the hair above and the excitement of eyes, nose, mouth, pink cheeks, ears, and face hair below. All these items appear more impressive or more charming, as the case may be, by reason of the calm open space afforded by the forehead. The classical forehead is high, wide, and unlined. This does not mean too high nor too wide. Good proportions are infallibly a golden mean.

So far as "unlined" goes, that is neither more nor less of anything, it just is. A clear forehead considered for itself alone is a very worthwhile feature to cultivate; when smooth and unlined, as with people who have peace in the heart, it can be a means of reassurance to friends.

The King of the Face Muscles—Scalp and Forehead

The muscle I am about to describe is totally unlike any other one in the face. It is big and flat and it blankets the whole top of the head, forehead included. As misunderstandings about this muscle abound, I shall start off by eliminating as many of the false notions as possible.

Attachments: This muscle does not grow out of the skull bones; it does not even clamp down on them the way barnacles do on rocks. It has only two attachments to the bone: one in front, to the top of the nasal bone, and the other, at the back, to the bony bumps at the base of the skull. In between these two bone attachments the muscle is entirely loose and flexible and slides freely over its bony base.*

The scalp is not exclusively a sod for the hair to grow in. If it were, it would stick to the ledge underneath like any respectable piece of sod. Why then does it have the capacity to slip around in this fashion?

---

*In strict anatomical terms this cap muscle is made up of a fore part and a rear section with connective tissues in between; but for purposes of face lifting it serves as one piece.

Because this strange scalp piece is a muscle: from eyebrows to nape of neck it makes a neatly tailored, close fitting cap. From now on I am going to call it the *skullcap muscle.* The anatomical fact that the skullcap starts—not at the hairline, where we imagine it would—but at the bottom of the forehead, beneath the eyebrows—turns out to be of outstanding importance for our purposes. (See page 45.)

What the Skullcap Does for Us

Direct function: Every muscle has its specific function, and every muscle operates by contraction; so does the skullcap. The direct function of the skullcap, operating by contraction, is to lift, stretch wide, and smooth out the forehead. Obviously this is extremely important; but the activity that establishes the cap as the first, though almost totally unappreciated, muscle of expression is its *indirect function,* as *partner* for the muscles of eyebrows, eyelids, and nose that lie along its lower border. You will not understand this aspect of the situation until I have had a chance to discuss it in relation to these other features. So we will await the proper moments for that and in the meantime continue our investigation of the skullcap for its interesting self alone.

Who Is Boss of the Skullcap?

Conscious and subconscious control: Do you remember from school days the trick one of the boys had in which he appeared to be suddenly moving his wig from the front to the back of his head? And he knew he could always get a laugh by wiggling his ears. For all that, he wasn't double-jointed or double-muscled or anything peculiar. He was just making some sort of use, however foolish, of a muscle that we all have but that few of us know how to make use of. You will enjoy the following quotation that comes from *Scientific American* for October, 1965. The title of the article is *The Origins of Facial Expression* and is written by Richard J. Andrews:

> In macaques and baboons *scalp retraction* (italics mine) is much exaggerated and produces a spectacular display, such as the flattening of the topnot of hair or the pulling up of folds of brightly colored skin from under the eyebrow, which gives the animal a look of staring astonishment. Like ear flattening these displays carry information, but they have no element of physical protection: Their solid function is communication.

My dictionary states that the macaques "are any of short tailed monkeys of Asia and East Indies." These little creatures can be a source of encouragement to us for helping get the scalp muscle going again! But now we return to genus homo sapiens where, except for the deliberate outcroppings of conscious control, the grand old scalp and forehead muscle has been going through the generations exclusively upon order of the unconscious. But surely you cannot expect the skull cap to put out the extra effort to hold the face up when all it is ever asked to do is to protect skull bones and to grow hair. This is the place for the conscious mind to intervene.

What Happens to the Skullcap as the Years Go By?

We have to see clearly what goes wrong for this feature (as for every other one we shall encounter in these pages) in order for us to get the dramatic impact on what not to do with it. With this kind of foundation the suggestions for right maintenance of the muscles will be vastly more telling and effective. Instead of being just little drills tried out for a while and then abandoned, these conscious activities will be taken on as a permanent part of the reasonable life where right functioning and beauty go hand in hand.

So what happens? For one thing, the hair changes its color and thins out; also the scowl wrinkles and other markings come to the forehead; also the cap elongates, causing little curtains over the eyes and a sagging nose. In fact, the sag in the skull cap is responsible for the initial let down of the whole face.

The other day I rescued a baby where I found him sitting propped up against the pillows in his carriage. His little loose-fitting cap had worked its way down over his eyes so that he could no longer see out. The more he tipped his head back to help himself see, the more the cap came down over his eyes—a terrible predicament! But this is actually what is happening to many of the older faces, with results just as displeasing to the wearer. Day by day our skull caps are coming down a fraction further, even to the point where drooping eyelids are partially obscuring our vision. Then, too, through the system of interlocking muscles, the sag in the cap muscle shows up as far down the face as the jowls and throat. Doing the right thing for the cap is consequently going to be an influence for good in every other feature of the face.

One day a woman came into a beauty studio in New York where I was a consultant. She was troubled by an acute, involuntary drooping of the upper eyelids and had to tip her head back in order to see.

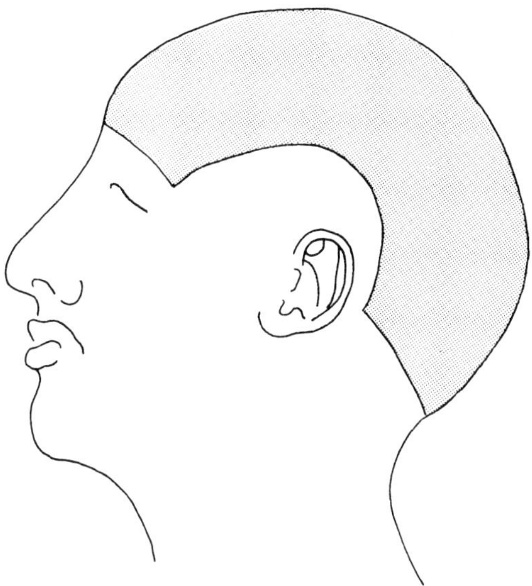

THE GREAT FOREHEAD AND SCALP MUSCLE*

Keep it high.
Keep it back out of your eyes.
Try to wear your little cap on the back of
   your head and watch it pull up your face!

Drooping Lids

The abandoned skull cap gives the whole face its initial let-down! The rumpled forehead. The dropping eyelids. The sagging nose.

*In strict anatomical terms this cap muscle is made up of a fore part and a rear section with connective tissues in between; but since it functions as one piece, we face lifters use it for what it is for our purposes—one cap.

She was on her way home from a visit to one of the leading plastic surgeons who had told her that this condition of the lids was a disease with a long name, and that there was nothing she or the doctors could do about it, the chances being that it would get worse. I decided that it would do her no harm if I explained how the scalp muscle worked as partner muscle for the eyelids. I told her that I believed she could regain control of the lids if she learned how to avail herself of the skullcap.

She was interested right away and gradually worked out what I suggested. She practiced opening her eyes wide, wide as she could, at regular intervals. She went around with eyebrows up. She worked with the scalp lock (as you will see how to do in a moment) and she recovered from that condition completely. It was as I had thought, partially anyway, that the skullcap had gone sluggish through lack of use.

The primary reason for the *why* of the sag is that we so rarely call upon this muscle to do its work. And the reason we don't do this is, surely, because we are not aware of what it could accomplish for us. There is not one of us who, knowing what the situation was, would not wish to do something about it. I propose now to introduce you to some simple activities that will help recapture the healthy functioning of the scalp-and-forehead-muscle. The proposition is to keep the skullcap short and thick and ironed out by using it.

Techniques for Activating the Sleeping Skullcap

(1) The Skullcap Lift
Glance up into the mirror and:
Raise your eyebrows.

Everyone can raise his eyebrows. By doing this action you have demonstrated to yourself that you already have control over the lower border of the skullcap—the forehead division. The next step is to:

Raise your eyebrows, and, this time see if you can do it
without creating any horizontal wrinkles in the forehead.

If you can do this it is wonderful because it means that you are already able to draw back the skullcap all in one piece as you should. It will do no good to push a wrinkled up forehead against a tight scalp. If this is what is happening in your case:

> Grasp a lock of hair at the top of your head—scalp lock—
> fingers on scalp if bald, and for a few moments
> move your whole scalp-and-forehead muscle gently back
> and forth by this means.
> Then practice drawing-back-the-lock-and-lifting-the-eye-
> brows, *as one movement.*

This will convince you, in practical experience, that scalp and forehead is all one piece.

Rest assured:

> If you keep eyes well open, eyebrows high, scowl muscles
> stretched out, forehead clear,

you are definitely on your way.

When you have learned how to use this muscle, all you have to do is to develop the habit of keeping it up, in tone, and short, as part of a posture habit. You will have to learn to wear the skullcap in jaunty style on the back of your head—not falling down over your eyes as though you had lost track of what was going on.

I trust the day will come when young people will learn early in life to develop control over this muscle; and those of us who are older may remember that an active muscle plumps up and shortens. To a heartening degree, faces that have already begun to droop may be lifted up again. The principle is the same for old and young.

(2) Scalp Massage and Nutrition

We will conclude now with two additional procedures that can by no means be overlooked. The first one: now and then as you feel the need you will do well to moisten the fingertips with a minute amount of lanolin (or more, according to the dryness of scalp) and for a few moments give the scalp a brisk massage. This will set the circulation going and aerate the roots of the hair. The hair will respond to such attention almost at once. It will stand up better and look more alive.

Beyond this surface ministration, another, and a deeper means to healthy, vigorous scalp and hair, lies in right food for the body as a whole. At this point in my text I turn you over to a professional

nutritionist, Adelle Davis. In her remarkable book *Let's Get Well**
Adelle reassures us of many things, amongst them that even graying
hair can be controlled, to a considerable extent, by right nutrition.

Birds Eye View in All Directions

We have found out that there is much we can do about this impressive feature, the dome of the head—about the muscle side of
the story anyway—the skullcap with its hair, its forehead, its eyebrows, and incidentally, its whole face. As for the bone side of the
story that is another matter. In heads like Tagore's we can appreciate
how many generations must have been at work sculpturing that
magnificent dome. In terms of the individual, we have to recognize
that there is little we can do about the shape of our heads. However,
the head-shape can do a great deal for us—that is, if we understand
it in design and make use of it as the foundation for the whole scheme
of our ensemble.

The Talisman of the Perfect Oval

The perfect oval, or egg shape, is a marvelous form; but, obviously,
few of us have any such wonderful shape in the way of a head. As
we know, the anthropologists have us all classified on the basis of
head shape: the long, the round, the in-between, the interracial
heads, not to mention all the individual variations upon these themes.
There are, in fact, as many head shapes as there are fruits, nuts and
vegetables. It can be helpful to go to the mirror and, pushing back
the hair, figure out which vegetable, fruit, or nut, this face of ours
most resembles. Is it a pear held stem up or stem down? Or a hazel
nut? Or an Idaho potato? If we can then draw this shape with a
three-inch vertical axis, and lay it over the perfect oval that we drew
earlier in the chapter (also three inches) we will see exactly wherein
our problem in design lies. Is the face too small at the chin, or too
heavy? Is it too narrow at the forehead or too wide? Does it peter
out equally at both the top and bottom? If it does any of these things,
then the approach to our design problem should be one of counter-

---

**Let's Get Well* by Adelle Davis, Copyright 1967, Harcourt Brace and World,
Chapter 12, *Skin Problems* (scalp) *Are More Than Skin Deep.* For *graying hair* see
close of chapter.

balance. If the oval is too flat on top then we must do all we can to fill out with the hair. Is it too wide? Then we see what we can do to set up a transverse or vertical direction with hair, hats, or necklines.

We all may live to see the elegant society leader at the reception whose gown, black, cut with a low V neck, exaggerates her long face and nose and brings harsh emphasis to her aging throat. Had she worn something soft at the throat with its lines judiciously disposed across the vertical, this would have helped to bring the basic egg into balance and so afford the beholder the visual freedom to enjoy immediately the delightful expressiveness of her face and her fine brow and hair. Then again, the opposite error: we may yet see the lady with the widest face in the world appear at her own song recital in a gown with shallow neckline cut from shoulder end to shoulder end. Of course, harmony is an essential ingredient in design —bringing emphasis through repeats in variation—but we must know what to harmonize with, and what to tone down by balancing out in other directions. By means of our talisman of the perfect oval, we can at least be assured of setting out in the right direction. We must not be too abrupt in the changes we are trying to effect, just gently suggest the lines best calculated to bring the whole into balance.

People in general seem to have a better instinct for bringing out a good feature than for minimizing a less good one. Take color, for instance. We do not often have to remind a man how he can bring out the color of his eyes with a touch of the same, or related color, in his tie. We know, too, how refreshing can be a contrasting color: blue-green or turquoise for the red-head; or the right shade of rose on a woman's lips which helps to bring out the blue of her eyes. But, actually, color should be looked upon as a last, though gloriously rewarding, touch to be thrown in after the more demanding problems of line and proportion are settled.

Head Shapes and Hairline

Occasionally we see a pretty girl who has lost some of the beauty and expressiveness of the forehead because a heavy crop of hair has grown too far down over it. I believe she would find herself well re-paid to have some of the surplus hair removed. I have also seen an attractive girl with too large a forehead, one that, were it all exposed, would give her a denuded look. This, for a woman, offers little problem as she can dress her hair down over her forehead as she pleases.

Men with a receding hairline will, I trust, remember that the additional forehead brings emphasis upon the intellect. We remember, too, that if professional necessity requires it toupees of marvelous craftsmanship are available. A man or a woman should feel as free to slip on a hair piece as he or she does to put on a hat; it has precisely the same purpose to protect and ornament the top of the head. All we need is candor and a sense of humor. Let us approach the matter like good actors—wear it or not wear it according to the role we are playing.

In any case, whatever alterations we may see fit to make in the hairline, we have always the classical model to turn to for the last word in good proportions. It is for us not to attempt to copy, but to make use of this standard in helping us figure out how best to handle our own problem.

Head Shapes and Hair Styling, Men and Women

The basis of becomingness in the art of hair styling lies in the arrangement that will help the face appear better proportioned. Whatever the type of face, the aim of any able stylist will be so to design the hair that the sum total of face and hair together may work toward the balanced form. Master stylists are those who find ways of adapting current fashions to this end. An occasional exception to the rule in no way upsets the general principle.

We do well to remember that it is one thing to indulge in the exaggerations of current fashion, knowing what we are doing and how much we can get away with; and it is quite another matter to be fashion's victim against our own best interests.

Men whose heads are not of standard shape know what terrible things can happen to them when they submit to a routine hair cut. I remember seeing one interesting long face cruelly cartooned by his barber who had given him a haircut close to the sides of the head and high on top. A man having difficulty finding a barber who will cut his hair the way he needs it, will do well to find one who cuts hair also for women. These men and women are likely to be more open-minded.

And now for a word about hair exclusively for men:

Face Hair and the Basic Egg

The clean-shaven face will forever hold its own, in or out of the fashion parade—Michael Angelo's *David* needs no adornment in a beard. However, the man who looks best in some combination of sideburns, moustache and beard should take full advantage of the fact that face hair is available; it is a truly splendid design opportunity.

The general rule for the shaping of beards is the same as for top-of-head hair, it should help draw the face pattern into a more harmonious or more challenging whole. It makes sense for a man to divert the eye of his beholder from a not so perfectly proportioned nose to some attractive sideburns, or help bring width to a long face with a smiling moustache, or tuck away an unimpressive chin in a handsome beard. Only genius, your own or that of your friends, will tell you when off-standard proportions are so uniquely interesting that they can afford to be exaggerated. The color, quality, and upkeep of the hair are, obviously, major considerations. How beautiful with some of the younger men is the color of the hair!

A special word now about beards for older men. Beards on younger men have a tendency to make them look older—not that that is important one way or the other if the design is good; but beards on older men, well designed, can be so especially right for them. These beards add a welcome touch of dignity and help to bring out the more mature personality.

Before concluding my remarks on face hair let me go back for a moment to the subject of the clean-shaven face. We can hardly forget that the *mouth,* for anyone at all, is often one of the most expressive, delightful and sometimes impressive of our features. To cover up such a feature with a mat of hair, no matter what its quality or color, can surely be reckoned as loss to the total effectiveness of the design. Then, too, as we have seen and shall see again, the flesh of the face all by itself is often lighted with so much sensitivity and freshness— "the golden mean"—that it hurts the eye of the beholder to have so much of it disappear from view behind a screen of hair. We can judge of this to some extent when we remember how it feels to talk with someone whose eyes are hidden behind dark glasses. We feel out of communication.

The Color of the Racial Egg

When we analyze skin color we find that there is no basic difference in the "color" of the different races. If anyone will analyze his skin for color he will discover that it is one mix or another of red and yellow with a touch of neutralizing blue. What confuses most people who have not had experience in color analysis comes from their not realizing that red and yellow show themselves to the eyes of the observer in all *values* from palest shades of pink and primrose to the deepest reds and browns.

How much the different racial shades are due to the different amounts, angle, and intensity of the sun's rays in the different latitudes, I am not equipped to say. But I do know that a white woman of great beauty, having on one of her travels acquired a deep suntan, was for a time taken for the queen of one of the exotic islands to the south. The employees of the boat on which she was traveling went scurrying in every direction to serve her. It would seem that if in the course of a few months, the sun could change a racial shade to such a degree as this, it might be reasonable to suppose that over archaeological periods it would be likely to develop a shade that would stick on. In any case the flesh tone of the various races is measured more by degrees of light and dark, than by a change of color. Skin is a transparent epidermis and blood is red the world over. How splendid amongst us are all the differing shades: pink, coffee-cream, white, bronze, ebony.

Standard and Off-Standard Proportions

As we experiment we find that there is only one type of face that looks well in any hair style, hair cut, hat, or neckline and that is the one whose proportions come closest to the standard. The rest of us will produce our own particular masterpiece through a judicious balancing out, or bringing emphasis, as the case may be.

I hope some day to have the fun of showing, on the screen, what can happen to a perfectly good face, somewhat off-standard in proportion, when it serves as model for a procession of hats in a wide variety of styles. The hats for this experiment should be new, untrimmed, and black. There should be no change of makeup or facial expression for the model, in order that the audience may concentrate their whole attention upon the effect that the lines of the hat produce upon the face. As these hat lines change, the face, to a most extraordinary degree, changes character. Its very proportions seem to shift. It may descend all the way from the top to the bottom of the social ladder and the reverse. But there are also some hat and face combinations that have nothing to do with top or bottom of anything but only with laughter for its own sake.*

---

*Could we but bring back Robert Benchley to dub in the titles!

DIAGRAM OF THE GOLDEN MEAN*

*This drawing is taken, by permission, from an article in *The Art Quarterly,* summer issue, 1962. Here, two inspired young artists, Donald Edward Gordon and Francis Cunningham, have figured out afresh the famous lost canon of proportion, "the golden mean" of Polykleitos. Their work was done from photographs of the statue, Diodoumenos, by this sculptor, and checked against a replica in the Metropolitan Museum of Art, New York City. This is a most fascinating and important article that is likely to be of interest to artists in any field of endeavor.

Head of Aphrodite. Style of Praxiteles. Greek, 4th century B.C. Parian Marble. Francis Bartlett donation 03.743. Courtesy Museum of Fine Arts, Boston.

Michelangelo's David: Accademia di Belle Arti. Florence, Italy. Done in the classical style.

Chapter 4

EYEBROWS AND NOSE

In some ways the nose is the most challenging feature we have. It comes first, physically, that much is certain. When well placed in the face it falls precisely within the center third of the oval, spang in the midst of everything. Nothing can deprive it of this splendid position of eminence. Sometimes when its proportions deviate too far from the standard it can cause a good deal of misery.

As for the eyebrows—it is possible that their original purpose was a functional one, to divert dust and moisture from the eyes; however this may be, the eyebrows are the purest of pure design, and one of our most lively and interesting features.

The chapter on the skullcap suggested the cross within the oval as the foundation for the face pattern. Now, by means of nose and eyebrows together we will see that linear pattern develop into an architectural one. You may remember from the first chapter that the face suggests an architectural structure of some sort—a little one man temple perhaps. The nose serves this interpretation as central column; the eyebrows as arches; the lines and forms of the mouth as base for the column of the nose, with the chin as the substructure. The most architectural of all face styles is the classical, so we will now analyze the classical standard for these two features together.

The Classical Nose

It is apparent that the nose which best serves the function of central pillar is the one that goes straight to work without stopping off for bulb, bump, or beak on the way—which again does not mean that we prefer this type of nose to all others. The classical, we remember, is a standard of reference, not an ideal for the individual. This standard of measurement was, obviously, the ideal for a certain people; but we have only to think of the great beauty of the oriental faces, and the Black, to know that it is not the standard of beauty for us all.

The classical, or "Grecian" nose has a high bridge that is almost a continuation of the plane of the forehead. The width of the nose—equal to the space between well-placed eyes—is equidistant throughout its length, neither tapering nor spreading as it goes down. The under plane, containing the apertures of the nostrils slopes neither up nor down, but is at exact right angles to the face.

Classical Eyebrows

Classical Eyebrows take their rise directly over the inner corners of well-spaced eyes, and proceeding along the ridge of the eyesocket, level over the eyes, drop away gently to their conclusion. Eyebrows come, obviously enough, in many different styles: the straight, the round, the pointed or Gothic, and all the many in-between types, each in its own way delightful. The classical, horizontal brow, more than any other types, contributes to the face the elements of strength, serenity and peace.

Usage

We saw something of the results of "time" upon these features in an earlier chapter. The nose, whatever its form may be, would seem to get more so with the years—more bulbous, more beaked, whatever; but the general trend for the nose is, I believe, to sag down, especially if it is a heavy one.

Eyebrows, too, tend to lose their animation and droop down. They also draw together in the familiar scowl wrinkles. Under the stress of our tensions these eyebrow muscles grow shorter, and eventually stay shortened all the time. Of all face blemishes due to improper use of muscles, the scowl wrinkles are the most obvious. There are many splendid older faces that have these lines; but such, let us not forget, is the case with every one of the aging face symptoms under consideration. If we think the matter through, we will recognize that the older faces are not more splendid because of these effects, but in spite of them. In the faces of young people these lines represent a feeling that will age them prematurely.

The Pullers-together of the Eyebrows

Two muscles control the eyebrows: The first is a diminutive *pair* of "wings" left and right just above the nose; their lower borders

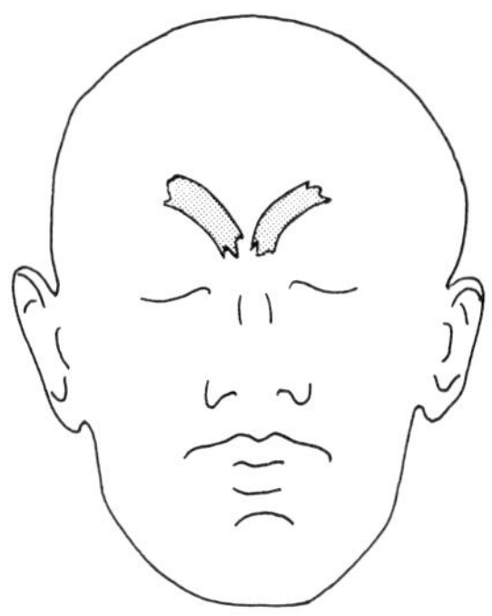

THE PULLERS-TOGETHER OF THE EYEBROWS—
"THE SCOWL MUSCLES"

Relax them! Let them go!
To counteract all the in-pitching forces, lift them by means of the skull cap
and, as you will discover later, by means of the ear muscles as well.

Eyebrows Falling Into Sockets

Scowling Eyebrows

Up Eyebrows

attach to the frontal bony ridge and, above, their tissues blend with those of the skullcap which covers them. This pair has but one function—to draw the eyebrows together in a frown.

Although we are well able, consciously, to control these muscles, the fact is that we do not very often do so. The scowl lines creep in upon us without our knowing anything about it and that is evidence that they take their orders from the subconscious.

The second muscle, a single one this time, which completes the full maneuverability for eyebrows is our old friend the skullcap. What a wonderful job the skullcap can do for the scowl wrinkles! It gives them their backward stretch. This is the first of the places referred to in the previous chapter where the skullcap serves as partner muscle for the features that connect with its lower border. In eyebrow muscles and skullcap we have a beautiful pair of muscles working opposite to each other: Eyebrow muscles contract—and the scowl appears; skullcap muscle contracts—and the scowl vanishes; pull-in-and-down, pull-up-and-out; pull, pull.

And now, the Nose

As makeup artists for stage and screen so well know, we count on the nose to tell sweeping things about the character: his race, for instance, or his long inherited position in society. A family can produce a certain type of nose and stick to it with remarkable tenacity through the generations. Beyond this framework the general feeling seems to be that the shape of the individual's nose has been bestowed upon him either in the lottery of chance or by the Creator's whim— that there it is, and that is that. We almost never take into consideration the extent to which we make our noses for ourselves.

The Unexpected Muscles of the Nose

There are eight major muscles for the nose. We will look first into the five little ones designed especially for expressiveness. One pair of these compresses the nostrils, one pair dilates them, and there is a solitary depressor on the end. Since the nasal bone extends only about a third of the length of the nose, the rest being stiffened by malleable cartilage, we can readily see how much the end of the nose is influenced by the individual who owns it. Nor is it hard to appreciate how the nasal form in its entirety can be molded, over the years, by the interior life of the passing generations. My mind turns to those characters of the Italian Renaissance that we see in the art of that period—their long noses with the compressed aristocratic beak. See Dante's nose.

After the muscles of expression we must not fail to look into the three small muscles that tie in directly with face lifting. The first of these is a long and slender one that extends the full length of the nose's top plane; its function (in conjunction with the scowl muscles) is to produce a particularly awful expression of rage. However, we can make good use of this one, too, for beauty, because of the fact that its tissues, at the top (as we have just seen was the case with the scowl muscles), are interwoven with those of the skullcap—the partner muscle designed to pull back the other way. When subjected by this potential face lifter to the light backward stretch, this small friend can act as checkrein to the nose, and so help keep this feature from sagging down as the years go by.

Lastly, there are two vertical muscles that blend the sides of the nose in with the cheek tissues. These operators give us an important tieup of nose, cheek and upper lip. By means of these, any impulse that lifts either the cheeks or the upper lip, lifts the nose also. So the force that lifts the nose here at the sides is the *smile* and all degrees of laughter from the first almost imperceptible lift of the upper lip to the great big belly laugh. Yes, it is the smile and the laugh that provide the nose with flying buttresses!

We have already seen how important the nose is in facial design, offering as it does, the central pillar; and, by means of the eyebrow arches, the support of the roof, or the dome above. It is reassuring now to have the complete story in terms of the muscles, too. Even in a muscular sense the nose is not stranded all alone in the middle of the face without a helping hand in any direction. On the contrary the nose is laced into the whole of (1) face expressiveness, and (2) face lifting, by means of a full complement of living muscles.

We have seen earlier in the chapter that the direct function of the eyebrow muscles is to pull the brows together, and just now we see that the direct function of the top-of-nose muscle, in conjunction with these eyebrow muscles, is to produce an expression of rage; so therefore, the best we face lifters can do with the two of these is to relax them, and go about our business in relation to them from another angle altogether. Now it is time to turn for help to the long neglected partner up above, the great skullcap.

*Setting to Work the Partner Muscle for Nose
and Eyebrow Muscles Together*

Here we have something we can actively, positively *do* to help
clear the appearance of the forehead between the eyes, and hold a
checkrein on the nose.

We know that the top-of-nose and eyebrow muscles, and we will
include here ahead of time the great round muscles of the eyelids,
are all riders on the coat tails of the powerful skullcap. So we must
get busy now and set the skullcap to work in the interests of all these
three together.

1. Grasp the scalp lock and move the cap slowly backward.
   Relax.
   Do this a few times.
2. Then, *as the cap draws back:*
   Let it lift and stretch wide the eyebrows
       stretch them wide
       stretch them wide;
   Let it open wide the eyes
       open them up
       open them up;
   Let it give the nose its upward lift
       lift it up
       lift it up.
3. And last but not least, especially for the nose,
   Remember the laughter—
       all the different kinds
           of wonderful laughter!

Yes, this big cap muscle up above can be held responsible for:

1. the up-eyebrows
   that we find in attractive, animated faces;
2. the well-open eyes
   the colorful, expressing eyes that we want to see all we
   can of;
3. the well-sustained nose
   preventing the unsightly wrinkles on the nasal bridge
   from growing on.

Respect for Eyebrows

I remember a young woman, a nurse by profession, who came into the cosmetic studio where I worked in New York. She was an exceptionally beautiful girl, with more classical features than anyone I have ever seen. But her eyebrows were foolish, inadequate affairs. She had, obviously, been plucking them. I suggested she let me draw in the whole eyebrow, the one that would seem to me to really belong to her, just as if I were making her up for stage.

"That is the way mine used to be!" she exclaimed as the new brow appeared. "I haven't had them like that since I was a girl. I'll bring you one of my childhood photographs to show you." It was not surprising to me that the forces which turned out such perfect proportions of face and feature and such a quality of the flesh had not failed her at the eyebrow.

I have also seen women who have plucked their eyebrows so that they had a slightly swayback effect, or festoon, which imparted to their attractive faces a curious quizzical expression.

The way to judge the rightness of an eyebrow is to search it to see how well it fulfills its architectural function. In conjunction with the nose does it appear to support the roof or dome above? If it does not, perhaps our understanding of the structural necessity will help us find the answer.

We cannot, of course, take this business of the arches too literally. The highest point in the true arch, surely, is in the middle; but in the eyebrow it is out past the center. This asymmetry helps to keep the facial pattern dynamic, helps move the observer's eye outward from the center.

Eyebrows are not just "lines," however fine these lines may be. They are forms which start off strongest at the nose and taper as they move out to their conclusion. And here follows a subtlety that few people except artists discover: the direction of the hairs presents a lovely fan-like movement—up, over, around and down—a movement that plays a definite role in the design, helping to relate eyebrows to the face as a whole. Notice, too, that the brow's upper edge is crisp and well defined. This is the part that makes the eyebrow "carry," but the lower edge is soft and helps blend the brow to the face.

Plucking

Eyebrows are so important to the face that they should not be tampered with by anyone who does not fully understand their place in the design. When proportions of face and feature are good, we limit the plucking to a clearing away of stray hairs. Since we aim for high eyebrows we should never, under any circumstances, interfere with the top edge. I have seen one young woman who, in her effort to thin out the line, had plucked down and down on this precious top edge until her eyebrows appeared to have almost fallen into her eye sockets—a major makeup calamity.

Most plucking should be done from below, which as it thins the line, serves also to "lift" the eyebrow. But even this much plucking should be done reluctantly. It is always followed by a loss of softness at the lower edge, and the design loses one of its more subtle values.

If the space between the eyes is too narrow, a dexterous removal of a few hairs at the inside ends of the brows may give the appearance of widening this space a little. This must not be overdone or the effect will be as if the face architecture were flying apart. Then again, if the nose is long, it may be unwise to clear away even the soft hairs between the brows, for these serve as a modifier to the length of the nose. Nature is a fantastically clever decorator, and if we are not perfectly sure of what we are doing, we had better let her have her way.

Eyebrows held aloft by means of the great scalp and forehead muscles are for everybody.

The Nose with Off-Standard Proportions

Plastic surgery is a deeply important branch of the surgeon's art—some of these doctors are equally surgeon and artist. Incredible rebuilding jobs have been done and are continuously effected in this field. I speak of surgery here for the following reason: We must make sure we don't ask the cosmetic branch of surgery to go too far. There are many ways, short of surgery, of helping along with a problem nose.

Make-up

I have seen one rather large nose dealt with to the satisfaction of all concerned by finding exactly the right shade of powder for it.

Let us imagine that we have a powder that is a perfect shade for the face as a whole. This powder should be used for all the face except the nose. For the nose we need one *fine* shade darker than this overall powder. Under this light veil of "shadow" the nose becomes less obtrusive, appears to fit in better with all the rest. In order to emphasize this point let me add: Never try to cover up a troublesome feature with a powder lighter than the flesh tone—it will, of course, have the reverse effect.

Powder should neither be smeared on, nor dusted on, nor rubbed in; it should be pat-pressed. It should lie on the skin's surface like the powder on the wings of a butterfly. We must remember, too, that powders well selected and well applied, do not show as powder on the face. They just leave the skin more dainty, fresh, and attractive, and the face, if the design has been well understood, at its best as to proportions.

One more point for women: The strengthening of an eyebrow, very delicately with a sharply pointed pencil can, on occasion, greatly assist the nose. Instead of weighing so heavily there all alone in the middle of the face, it can, by means of the eyebrow-arch appear to be linked firmly to the side walls of the head. A little design trick for the initiated.

Hair Styles, Hats, and Necklines

Obviously, fashions in hair styling and costume change so fast that it would be impossible for me to make specific suggestions. However a few generalizations may prove useful. Sometimes, hypnotized by our own lines and bullied by current fashions we find ourselves doing exactly the wrong thing. First, about the hair:

When the hair is dressed flat on the top of the head, parted or straight back—so beautiful for the face with standard proportions—it can give no support at all to a nose that is heavier than we could wish. A woman can help bring the nose into balance with a particular hair style. A man's haircut can help or hinder in exactly the same way. A good shock of heavy hair on the top of his head brings support to a large nose. Face hair, too, rightly designed, gives the man additional recourse. The whole style of him helps or hinders—his posture, his walk, his expressions.

Now a word about hats. Imagine, if you will, a hat shaped like an inverted bowl descending over a bell-shaped nose. Clearly, such a nose needs counterbalance. At least some of the forehead should be in evidence and the brim of the hat should offer some transverse lift. The reverse is equally true. I have seen a fly-away hat perched above an exaggeratedly tip-tilted nose. It is easy to realize in the abstract that such a nose needs some airy shelter from the lines of the hair and hat above it—always remembering that we can defeat our purposes entirely with abrupt contrast. Easy does it.

One more hat: This one is not so much an example of dealing with an off-standard face or feature as it is designing for an off-standard figure—which, obviously, includes the face. I once saw a very stout lady with a very sweet face. This face was partially eclipsed by a hat (a lovely one in itself), white, with a wide brim shaped like an inverted saucer—and around the crown a wreath of waterlilies—a case, surely, of harmonizing with the wrong thing. "Sometimes hypnotized by our own lines. . . ." Of course the hat, not to say the whole costume, should have been balanced out with at least some lines suggesting the vertical.

This brings us to a further consideration of *necklines.*

The edge of the costume at the throat is, obviously, the "neckline." Whether this "line" produces the V neck, or the round, or the square —wide, shallow, deep, whatever—the selection of it can have an almost startling effect for good or ill, "becoming" or "unbecoming," upon the face. To vivify your thoughts about the relation of neckline area to the face, see Mona Lisa, (p. 118).

If a particular face, with its nose, is overlong or overwide, we can *counterbalance* with a neckline that suggests the opposite direction: as for instance, a generous V neck for an overwide face—the V draws the pattern down.

If a particular face, with its nose, is overlong, and yet so distinguished that on occasion we might even wish to emphasize it, we can achieve this emphasis through *repetition,* or echoing of its proportions by means of the neckline area. Through experimentation you will find out which is the exact right one. (Hold pieces of colored paper cut in a variety of geometric shapes at the base of your throat!)

If a particular face, with its nose, has balanced, or "standard" proportions, the individual belonging to it can wear, successfully, any type of neckline at all.

Sometimes, perhaps, you have had a costume that you were aware was particularly becoming, and yet you did not know why, or consequently know what to repeat. The answer may well have been right here in a neckline area that balanced the proportions of your face. Watch for these considerations in the throatlines of those about you, and on television. If you are at all interested in design this is a fascinating place to watch its workings.

In the selection of costume some bad mistakes are made even in high places. We cannot count on the sales personnel to have all the answers for us. After all, their job is primarily to sell. We have to understand our own design requirements for ourselves—not forgetting that these needs change as face and figure change.

Although these last paragraphs on costume may appear to be of interest chiefly to women, I am convinced that men have picked up something of interest to them, too. Sound design principles are as relevant to the problem and opportunities of men's costume as they are to women's.

But, clearly, the subject of first importance to everybody in this chapter is finding out how to put to work the great scalp and forehead muscle in the interests of nose, brows, and as we shall soon see now, the eyes—all to help move on its way the lifted face. We are not likely to forget that the face muscles are the costume of the face bones, and (more than anything else,) the way we wear *them* tells the observer who we really are.

Chapter 5

THE EYES

Eyes are our gems without price. Nowhere in our physical world, so it seems, do tangible and intangible forces come into more absorbing interplay.

Not only by reason of their great expressiveness, but because of their position in the design, the eyes could not fail to dominate the face; in that important place just above center they gleam forth like lamps in the arches.

Classical eyes are large, set the width of an eye apart. The lids are neither overtensed, nor drooping—just well open. They are framed by the peaceful horizontal brow. Note that the plane in which the eyes are set slopes gently back from the forehead to the cheeks. This backward slope makes the eyes appear "well set," neither too deeply hidden in the sockets nor too far forward or bulging. Except for the stage, with all its intense, artificial lighting, well set eyes need no shadow from the makeup box. Sheltered as they are beneath the overhang of the forehead they are provided with their own translucent shadow. This shadow is further deepened over the ball itself by the projecting thickness of the flesh of the upper lid, and then again by the spread of the lashes.

The eye setting is, of course, partly in the interest of protection; but if the scheme of form, line, color, light, and dark had been designed exclusively for the purpose of enhancing the expressiveness and brilliance of the eyes, it could not have been achieved in more masterly fashion.

The Mysterious Round Muscles of the Eyelids

We look now with X-ray eyes beneath the skin's surface at the eyelid muscles, those bits of nature's engineering that contribute so much, over the years, to the eye appearance. To look at these little muscles in the dissecting room, no one of us in our wildest imagination would guess the heights of expressiveness to which they can, and do, rise in daily usage. Be their owner kind or cruel, honest or dishonest—whatever it may be—those experienced enough, or knowing enough, can detect the story in the flesh of the lids.

There is no question, of course, that the eyeballs have a magic all their own, and one that will perhaps remain forever unfathomable to the human mind; but when I speak of the talking eyes, remember that it is the whole area of eyeballs and lids together, from eyebrows to base of "tired circles," that does the talking.

These muscles include a larger area than their name, the muscles of the eyelids, would suggest. Each is in the form of a miniature doughtnut and extends over the same area that I have just been outlining as part of the eye in design, the whole area of the sockets.

The outer borders are attached all the way around to the bony ridges surrounding the sockets. The inner borders are exquisitely tailored to form the lids. When the eyes are closed these muscles completely cover the eye sockets. But it is important to note that these little muscles have more than bone attachments. All the way around the outer edges of the doughnut they blend their tissues with the other muscles that converge upon the eye sockets from the other directions. The special "other one" of these muscles for the lids is the great scalp and forehead muscle, the skullcap. The top edges of the eyelid muscles and the lower border of the skullcap are joined together; they are partners. Here we have at last the third of the features that ride the coattails of the powerful skullcap and are benefited by it—eyebrows, nose, and now, the eyes—more specifically the great round muscles of the lids.

The function of the lids, as of all muscles, is to tense and to relax. The varying degrees of tension serve the needs of the eyes from just closing, to squinting, to violent contraction of the lids. So here we have the partners ready to go to work in the interests of the eyes: Eyelids pull shut, skullcap pulls open; pull shut, pull open; pull, pull.

Let us look back now to the drill in the preceding chapter *For Setting to Work the Partner Muscle for Nose and Eyebrow Muscles Together.* I included the eyelids in that drill, do you remember? Now that you have the full story on these eyelid muscles, I urge you to run through this drill again, it is always refreshing (page 60). Then we will be ready to go on and discuss who controls these eyelids.

Conscious control: The eyelids are entirely responsive to voluntary

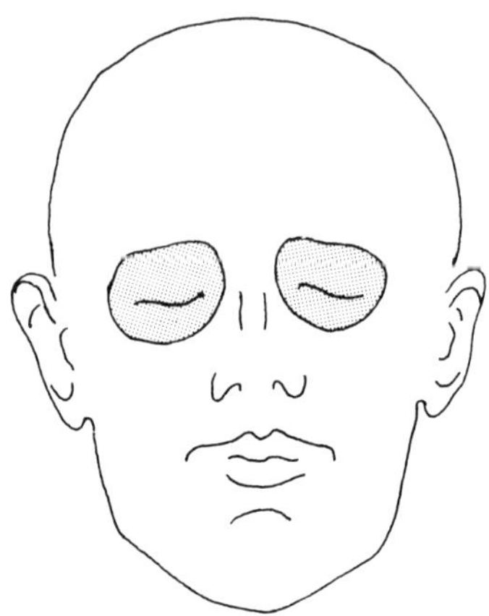

THE ROUND MUSCLES OF THE EYELIDS

Relax them! Never squint!

For exercise: Lift the upper lids and open the eyes wide, often and often, throughout the day.

For posture: Make this lifting a habit.

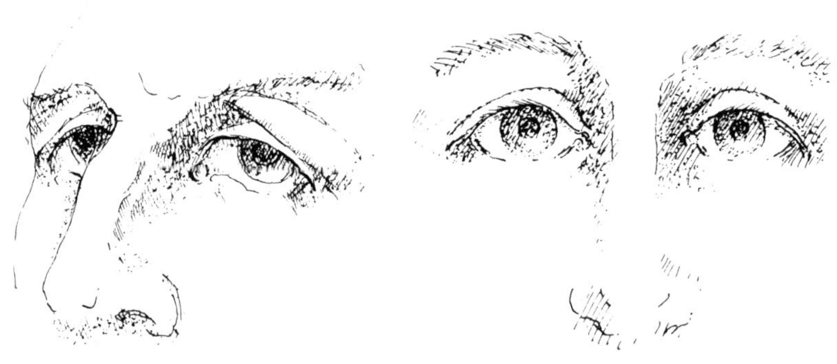

Curtains Over Eyes   Eyes Open Wide

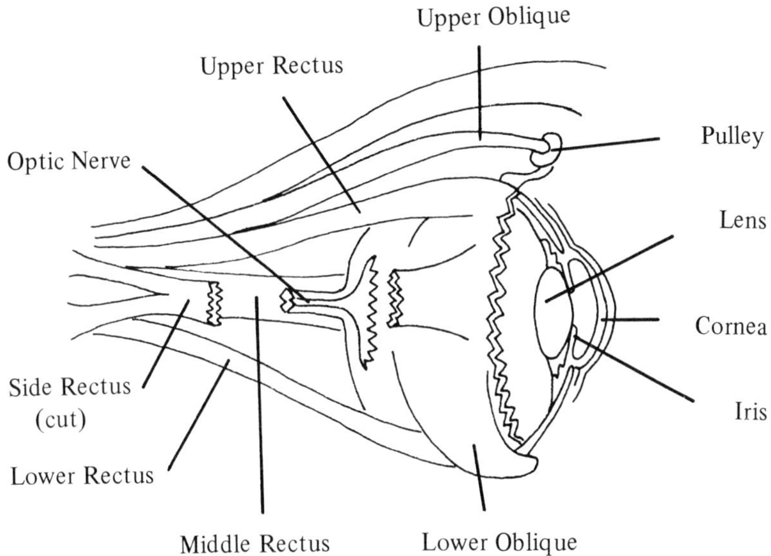

THE MARVELLOUS MUSCLES OF VISION

Relax them! Never strain them!

For relaxation drills. Do the blinking, sunning, palming and deep breathing
(you will see how shortly).

Eyes live on light:

From earliest student days have a good reading light

For all fine work, have a good light.

Have a good light!

control. Someone says, "shut your eyes," and we shut them; "open your eyes," and we open them.

Subconscious control: But for the most part the opening and closing of the eyes, and especially the manner of holding them open or closed, by day or by night, is almost exclusively in the control of the subconscious. In other words, we are most of the time just as unaware of what we are doing with these muscles as we are of the muscles of vision. We wouldn't be if we understood what a more voluntary control could mean in preventing the appearances we so rashly clump together under the heading of "old age."

Our Hidden Muscles of Vision

There are three major pairs of muscles for each eyeball. They are not inside the ball as you would somehow expect. They are within the eye socket but on the outside of the eyeball, and they control its movement. (See illustration on page 69.)

The first two of these pairs, the *Rectii*, reach from the back of the socket and attach themselves to the ball—one pair to the top and bottom and the other pair to the sides; their function is to roll the ball up and down and side to side. Look up, look down, pull, pull; look right, look left, pull, pull.

The third pair, the *Obliques*, also reach from the back of the socket and fastens to the ball; but their method of control is not so simple. They girdle the ball diagonally and, passing through a loop at the top, they operate from the back, pulley fashion. The Rectii and the Obliques, between them, enable the ball to perform the full movement of rotation achievable to the human eye. The diagram shows, roughly, how this is accomplished. The movement is, in part only, a voluntary one: We wish, consciously, to roll the eyes all around and we do so. Most of the time, however, they are responding to the dictates of the subconscious.

What further relation these muscles have to the actual process of seeing I shall not here attempt to say. The interior mechanics of vision is complex past belief (the ways of the camera are the ways of the eye) and should be discussed by professionals in the field. But I have said enough to establish the fact that the function of the eye muscles proceeds just as it does with the other muscles of the body: One set pulls in one direction and another set pulls in the opposite direction; pull over, pull back, pull, pull.

What Happens to the Appearance of the Eyes as We Grow Older?

Modern civilization has placed a terrible burden upon the organs of vision. They tend to fall into patterns of nearsightedness or farsightedness. They lose their liveliness and sparkle. Very often the lids pinch up around the ball making the eyes look smaller.

We are likely to see, also, manifestations of an entirely different nature. A little curtain comes stealing down from above, over the upper lid, sometimes covering the outer corners of the eyes and even part of the eyes themselves. This little curtain is a charming characteristic of oriental eyes from childhood on; but for occidentals it is another matter. We notice, too, reluctantly, the sag of the flesh beneath the eyes—the tired circles, or pouches. (footnote p. 74).

So we are able to distinguish in the muscles of the lids the results of both supertension and overrelaxation. In the following paragraphs I shall deal with each of these manifestations separately, beginning with the overtensions.

Excessive tensions are caused by a wide variety of usage: all the occupational tensions, such as much reading, or other fine work, responsibility for people or machines; mental effort; psychological or physical suffering, on and on; but the greatest offender in this respect is *fear*, all degrees of fear from the lightest anxiety, through outright fear: fear of losing what we have; fear of old age, disease, and death; fear of the unknown; even a light suspicion starts us peeping out through narrowed lids.

Conscientiousness, a socially acceptable form of fear, is responsible for a great deal of trouble in this respect. Tension in and about the eyes begins in the classroom more often than not. We can hope that the more creative education will bring relief to this situation. Amongst the new ways, children will be taught how to take care of their eyes, how to use and how to rest them, and, most important, what constitutes a good light to study by.

We will now look into some little drills devised by Dr. William H. Bates, M.D., pioneer ophthalmologist and surgeon, and designed to help bring relaxation to the over-tense muscles of the eyes and lids together. These techniques, easily demonstrable and tested over many years, are based on Dr. Bates's famous premise that (disease aside) *relaxed eyes see well.** Relaxing the eyes is not so simple as relaxing the biceps, so let us look into this important matter with care.

For Relief of Supertension in Eyes and Lids

1. Frequent and Easy Blinking

This should be the lightest and most airy flutter of the lids—five to ten blinks at a time. By spreading the secretions from the tear ducts over the ball blinking helps to lubricate and disinfect the eyes. It also helps to relax the lids and thereby assists in averting, or breaking up, the tensions within the muscles of vision.

2. Taking the Healing Sun on the Eyes

We all know from our own experience that sunlight, *taken in the proper amount,* revivifies and relaxes both at the same time. But you must never forget that sunlight, in excess, not only undoes all the good but produces an actual poison in the system. As a fine old doctor once pointed out to me: "In medicine the difference between a remedy and a poison is in the size of the dose." It is also important to remember that the rays of the early morning sun contain more life-giving elements than at any other time during the day. Noonday sun, pure heat, is not good for the eyes.

Take up your stand, then, outdoors if possible or at an open window but in the full light of the sun. If you wear glasses take them off for the moment, close your eyes and expose your face to the sun; with head held straight rotate your face slowly from side to side.

---

*Anyone wishing to study more deeply the method of Dr. William H. Bates, M.D. should read: *Quick Guide to Better Vision* by Margaret Darst Corbett, copyright 1957 byPrentice-Hall, Englewood Cliffs, New Jersey. Mrs. Corbett was Dr. Bates's outstanding pupil. She developed his method and taught it for many years in her *School of Eye Education* in Los Angeles. Aldous Huxley who benefited greatly by this method and wrote a book of his own on the subject—*The Art Of Seeing* (Harper) said of Mrs. Corbett's book: "Here is a thoroughly practical little book by the world's most accomplished teacher of the art of seeing." (Permission to use ideas from Mrs. Corbett's book has been granted by Prentice-Hall.)

Since staring eyes are tense, keep your eyes active beneath the closed lids by visualizing some happy scenes of activity—children swinging, a diving exhibition, trees blowing in the wind. The seeing mechanisms respond to visualizing with eyes closed just as they do with eyes open.

It is good to start doing this drill for one full minute at a time and work up gradually to three. Follow with twice as many minutes of *palming*. See paragraphs on *palming*, next in order.

Taking the healing sun on the eyes and face proves to be one of the most delightful of our face exercises. We know from experience that heartaches and tensions survive with difficulty in the face of the sun. Here we have at our disposal a moment of revitalizing that can be taken several times a day. So let us put away our dark glasses and allow the healing sun to bring up the sparkle of our eyes.

3. Palming the Eyes

To do the palming, rest the elbows on a table, cushion under elbows, so that the back will be relaxed and free, not crouched. Then, without touching the lids—cup the closed eyes with the palms of the hands. If you insert the fatty pads of the palms that lie just at the base of the little fingers into the inner corners of the eye sockets, letting the fingers cross lightly on the forehead, you can cut out all the light from the eyes. It takes a little practice, so don't be impatient. Do not, however, touch the eyes with the palms, to say nothing of pressing upon them. In order, at this time, to keep the eyes from staring, visualize in the mind's eye—just as you did for the sunning—some scene of happy activity. The combination of the restful darkness, and the healing warmth and other life-giving values of one's own hands, will be found to have relaxing value.

The periods of palming should follow the periods of sunning as the night the day. Palming is to sunning as rest is to activity. It gives the system a chance to absorb the good and elminate the waste that has resulted from the sun's stimulation. Palming periods should be twice as long as the sunning periods. Two minutes of sun to four of palming is good for a start.

When we do such drills as these to relax the eyes, we at once notice release from tension also in mind and heart. We are more accustomed to thinking of this process in reverse; relaxing of mind and heart helping to release body tensions; but remember that the process works in both directions and that the two together make up a cycle.

An example of the theory that relaxed eyes see well was vividly brought to my own door in the following experience.

One day I was having a reading session on purpose to improve my vision. I had been making use of the relaxation drills that we have been considering here. The print was not very clear to me, but I enjoyed the story so much that it helped me to keep on. The book, I remember well, was *The Laughter of My Father* by Carlos Bulosan.

All at once I came across a passage so deep in human understanding and so full of whimsicality and humor that I let out a great big laugh, as I did so, momentarily looking away from the page. When my eyes returned to the page, to my utter amazement and delight, the print had become intensely black and clear. I read on for a good two minutes with perfect vision. The sensation around my eyes was as though the storm had departed and the April sun at last was shining through.

Though the release from visual impairment did not last, still it left me knowing that my eyes were ready to produce good vision when I learned how to use them correctly.

For Relief of Overrelaxation or Sag in the Eyelid Muscles

The evidences of overrelaxation, excessive letting go are: the little curtain stealing down over the upper lid, and the sag of the flesh beneath the eyes, the tired circles. These manifestations are brought on by a variety of things: amongst them, fatigue and disease, even the pull of gravity. Gravity pulls down the lid muscles just as surely as it does every point of our unresisting flesh. Added to these familiar causes we have the negative emotions. In an earlier chapter I used the word *gloom* to cover these—all the giving up, apathy, despair. In the grip of negative emotions we throw in the sponge and gravity takes over.

I am now going to suggest a somewhat elusive little exercise aimed at direct stimulation of the sagging flesh beneath the eye—the lower section of the round muscle of the eyelids. It will not help you conspicuously if what you need is to go to bed early for a change, or have more exercise in the fresh air, or check what you eat;* but it is comforting to know that there is a way to order up this bit of flesh under the eye for the morning drill.

---

*For suggestions on diet in relation to eyes see Adelle Davis' *Let's Get Well* (New York: Harcourt, Brace and World, Inc., 1965), Chapter 29, *Eye Problems* (see Index also).

The Functional Wink

Take your hand mirror; smooth out your forehead and open your eyes well. Then practice a few winks first on one side and then on the other. *Do not squint.* The upper lid should remain wide open and stationary. Keep at it, you will get the feel of the action more and more and make bigger and better winks. After a while you will be able to activate both sides at the same time. Then, you will find that you can involve the face tissues all the way to the bottom of the upper lip, and even further, to the jowls and the bottom of the face. These lower tissues are given a form of passive exercise which helps set the circulation going in them, too. But let us remember—the specific exercise for the tired circles area is the *functional wink.*

We Return to the Great Skullcap

Here we come back again to the place where we were last discussing the triumvirate that ride the coattails of that scalp muscle and are lifted by it—nose, eyebrows and eyelids. So let us run over the scalp lift exercise again, this time, especially, for the muscles of the lids.

Use your hand mirror again.
1. Lift your eyebrows high
2. Lift up and iron out your forehead
3. Open your eyes wide
   stretch them wide open
   stretch open the crow's feet
   even feel the sag under the eyes lifting a little
   again and again lift your eyebrows high
   stretch your eyes wide open—
   open them up
   open them up
   open them up!

That feels good, doesn't it—like stretching the back after you have been bending over your work all day. In the interest of counterbalancing the list of causes that, on the one hand, tend to excessive pinching up, and on the other, to excessive letting go—we should avail ourselves of this drill often and often throughout the day for the rest of our lives. In one action it lifts the eyebrows, clears the forehead of the scowl, sustains the nose, and opens the eyes wide. In a most remarkable way it brings relaxation and stimulation, both, to these important face muscles, and all in the same action!

Eye Memo

When you are reading, especially if you read a great deal, now and then palm your eyes and then open them wide: finish off with five blinks. These two actions will help prevent eye strain. When you are intent over your work, whatever it is, develop the habit of keeping the eyes well open and the eyebrows high. While you are doing the dishes stretch the eyes wide open. Put a mirror over the sink; it will give you a laugh now and then. When you pass an open window in the sun, stop a moment, close your eyes, and turn your head in the healing rays. When you get into a bus and have a few moments to breathe, do a few light and airy blinks and stretch your eyes wide open. When you are in the garden or at the beach, don't let the sun bake you and your eyes to a crisp; but now and then, in the earlier or later times of day, face the sun, close your lids, and turn your head slowly from side to side.

During all the activities outlined in this chapter you are learning how to bring relaxation and stimulation—directly—to the great round muscles of the eyelids; and—indirectly—to the .eyes themselves. Once you have, consciously, trained the eyelids you can hand them back to the subconscious which, reminded now and then by the conscious, will preserve the good posture, and much more comfortably than it formerly did the damaging one.

Chapter 6

THE BOW OF THE MOUTH AND THE
SEVENTEEN STRINGS

The mouth cannot hope to rival the eyes in expressiveness. Take eyes of love; only eyes can produce such miracles as this. But for expressing the everyday comings and goings of life, the mouth has no peer.

We see a certain mouth grin, and we understand all the naive friendliness in the heart of the person who belongs to it. We see another mouth hanging open and marvel that such a state of vacuum can exist in the human species; we see still another gathering itself up like a rosebud; we see another shut like a steel trap. The experiencing in all of them could not be more lucid. The mouth is by far the most self-revealing of our features, and there are splendid, easy ways for us, consciously, to help sustain the mouth of beauty.

The mouth, ideally, has for the lips just as precise a form as the eyelid muscle has for the eyes; but the mouth muscle has a much less exacting function, and its attachments enable it to wander over a much greater area. The result is that the mouth, to an extent not shared by any other feature, is at the beck and call of the personality. All the other features can, to a certain extent, be considered as inherited; but the mouth gives away the whole game. By the time a mouth's owner is sixty, its expression is all his or her very own.

It is the homemade aspect of the mouth that makes it the favorite, and at the same time the most baffling, feature for the portrait painter. It remains hardly two seconds the same. For a dozen artists who can find a characteristic expressiveness for the eyes, only two or three can do equally well for the mouth.

I have on my desk some beautiful Japanese prints of actors "emoting." We see the actors in every conceivable grimace and we think of this face-making as traditional Japanese acting. And yet, if we take the time to notice faces, we can find many of our fellow citizens with the mouth in every bit as weird a contortion. For instance, here is a photograph of a gentleman whose picture was taken as he listened to the proceedings of some governmental agency in Washington. The mouth can only be described as horrible, a frightener-of-children. And yet I feel sure nobody noticed him—we are all too accustomed to seeing people just living their lives.

Usage, or What We Do to the Mouth Muscle

Such a variety of things happens to mouths as they are drawn this way and that by the action of the personality that it is hard to generalize about what happens to them as the years go by. I shall fall back upon some remarks made long ago by one of my art teachers. This man called our attention to the fact that in the course of a lifetime, the mouth completely reverses its direction. After a moment of incredulity we almost have to agree with him: there the mouth is, in the baby, squeezed between two fat cheeks into an indisputably up and down pattern. After a while it widens out so that it is about as wide as it is high—adorable. Later on it widens out some more and is charming indeed. Gradually it widens out a little too far, maybe, and then it widens out too far, definitely.

The other day when I was having my hair done in a beauty shop, I witnessed the following drama. The lady of this particular mouth had been giving the operator plenty of trouble which I had overheard. But now there she was at last all tucked under the dryer, its large silver bell well down over her head. She looked like a highly dissatisfied Roman general. Someone must have given him a clout on the helmet because there it was, effecting an all but total eclipse of his physiognomy. But the mouth was still in evidence, and it looked like the cutting edge of a scimitar. The line of the mouth, for line it was, could not possibly have been the mouth of current dissatisfaction only. It had been moulded by years and years of unfriendly spirit. If our warrior could have seen, on the screen perhaps, the picture she presented there, not realizing who it was, she would have laughed; knowing, she would have wept.

Do you find the thought a little terrifying, as I do, that every expression that flits across our faces, is qualified to a certain extent by every other one that has passed by that way since we were born? It is true though. More often than not the look on our faces is a composite of every thought and emotion we have ever experienced.

This situation, sometimes so tragic, is not inevitable, however, either psychologically or physically.

The Classical Mouth

The place of the mouth in the well proportioned face is at one third of the distance from the base of the nose to the bottom of the chin. Practicing artists know that the mouth is not built upon a flat surface but follows the curve of the underlying bone, its center being forward of its corners. It is this curve that is referred to as the mouth's *bow*. The form of the lips partakes of no extremes, they are full but not too full. The red parts (the width of the mouth muscle at the opening) are in good proportion to each other, and to the length of lips. When we visualize the pattern of the head and face as a little temple, we see that the mouth provides a base to the column of the nose, the length of the upper lip being the transition element.

Just as in the case of the eyes, there is no mouth without an expression. Even the "expressionless" mouth turns out to be a vivid description of a particular mouth. So the classical mouth has its own expressiveness. It is neither lazy, nor depressed, nor over-tense; it is best described, perhaps, as serene. Serenity, muscularly speaking, is activity and rest in perfect balance.

No description of the ideal mouth would be complete without a reference to the teeth. The way the teeth flash as the lips move over them helps the mouth to enter into competition with the eyes. Mouth and eyes are like actors, "stars," constantly vying with each other for the attention of the audience.

How the Mouth Is Built

The round mouth muscle is exactly the same kind of muscle as that of the eyelids, a doughnut shaped *sphincter* (one that draws in about an opening). The top of this "doughnut" attaches to the upper jaw bone just under the nose, and the lower attachment is to the lower jaw bone at the top of the chin. Understanding the mouth muscle helps us to interpret what the whole mouth pattern and expressiveness consists in: not just the red-of-lips, but the length of the lips, too; in other words, the whole circle of the muscle from base of nose to top of chin.

How the Mouth Muscle Operates

This doughnut shaped sphincter closes the lips. Light tension closes them; strong tension puckers them as in whistling and kissing. The lips help to manipulate the food in the mouth and help form the vowels and consonants in speaking and singing. Since this muscle has but one function—to pull in—it becomes apparent that a large share of mouth expressiveness must come from some other source.

So we turn now to the mouth's partner muscles. There are no less than fifteen muscles that pull back the other way—the cheek muscles. Cheeks and mouth build each other back and forth; they are but one expressiveness.

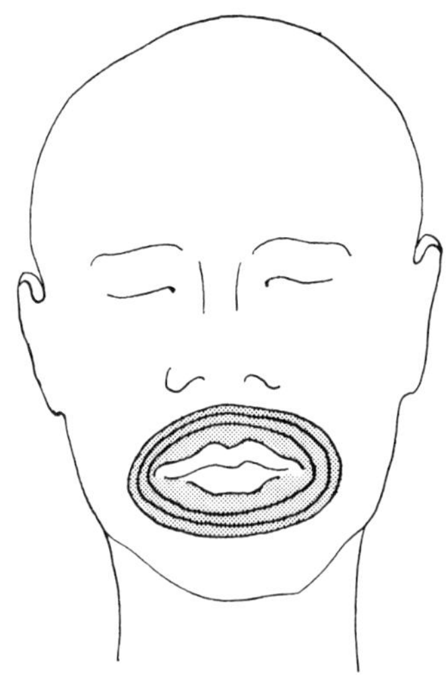

THE ROUND MUSCLE OF THE MOUTH

1. Keep it free from tension.
2. For exercise—keep whistling.
3. For posture (as you will see in a moment) keep it lightly lifted all the time by means of the cheek muscles. Get that little Mona Lisa smile.

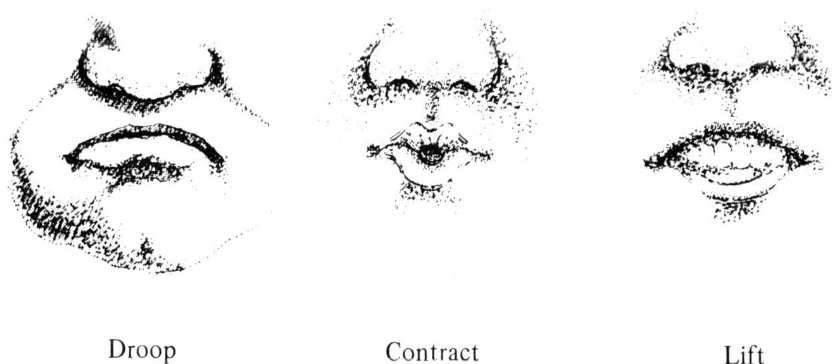

Droop   Contract   Lift

The Hidden Magic of the Cheeks

The first two cheek muscles are already known to us— the ones that lie at the vertical boundaries of the nose. They are a little pair, and they effect the central tie-up of nose, cheek and upper lip.

Proceeding from there on around the circle we come next to the strong smile and laugh muscles. We are generously supplied with these, two sets on each side, four in all. They make diagonal straps from mouth to cheek bones.

Then come the lifters of the corners of the mouth.

Beyond these again, pulling straight out to the side, we have the grin muscles.

Around the corners on the under side, come the draggers down of the corners, and then: the pullers down of the lower lip.

Lastly, right in the center, we have the only single muscle, the lifter of the chin. This single muscle attaches above, to the lower border of the doughnut of the mouth, then slips over the chin bone and mingles its tissues with those of the throat. Its function is to thrust the lower lip out. This is a good protest activity for little children. For adults, in these days, the action is not popular and the emotion it expresses is likely to be repressed. However, I have seen it among the photographs of one of our imperious thugs of the not so distant past.

This completes the cheek muscles, fifteen of them, standing ready to tug on the mouth. In a study of this nature we cannot over-emphasize that the upper group of these pullers-on-the-mouth are the form retainers and beauty builders of mouth and cheeks as well. It is a pity that we make such insufficient use of them; they should come in for double action if only in the interest of counterbalancing gravity, not to mention all the down dragging emotional factors.

It will throw light upon the similarly constructed round muscles of lips and eyelids to look for a moment at the fact that these important sphincter muscles do opposite things as the years go by. The eyelids tend to pull in, and the mouth tends to pull out. Why? The question leads us to search out once more the back pull in each case. For the eyelids the back pull is provided chiefly by the sleeping giant of the scalp-and-forehead muscle; for the lips this pull

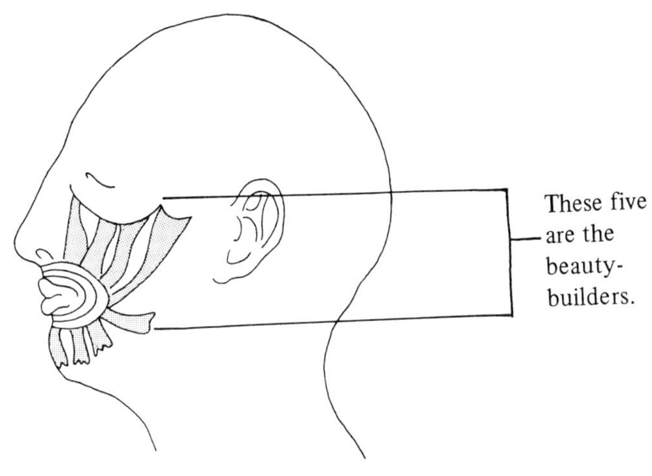

THE MANIPULATORS OF THE MOUTH

1. For exercise—tense them up as in a big smile.
2. To counteract gravity—keep them lightly in action the whole time.
3. See if you can whistle right along with the other elements of the lifted face. Pretend you are looking into the faces of children as you entertain them on your flute.

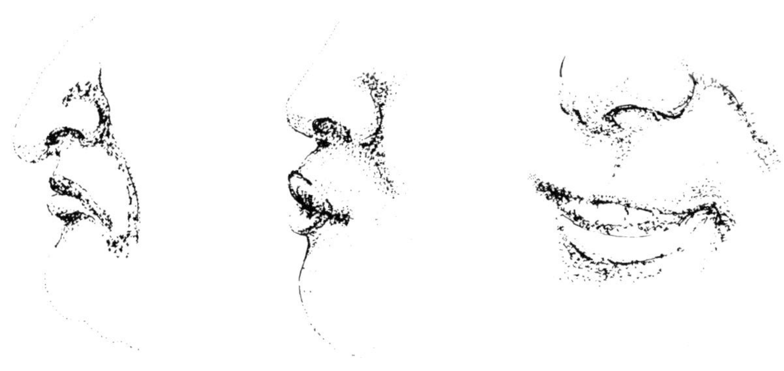

is provided by fifteen muscles, activated by all our positive and negative thinking with its happiness and sorrow, tears and laughter.

Opposite tendencies demand opposite treatment.

Techniques for Sustaining the Form of the Mouth

The Workout of Whistling

The one and only activity possible to the mouth muscle, considered by itself alone, is to contract. Whistling is a straightforward activity that gives this muscle a chance to limber up its own function. People who haven't whistled for years will be surprised to find how stiff, inept, and insensitive their lips have become; whistling will get them going again.

The Creating Smile

Now, try on a smile. The fat cheek that comes when we smile is due to the bunching up of a little pad of fat situated just below the skin as the muscles contract beneath it. As you make this awful grin you are probably thinking in terms of pushing out the mouth. But remember that no muscle pushes. Think instead of this action as being accomplished by the contracting of the smile straps diagonally across the cheeks.

The muscular results of smiling are to shorten and thicken: (1) the pair of muscles that lift the upper lip and sides of nose together; (2) the two pairs of smile straps; (3) the two lifters of corners of mouth; and (4) the two grin muscles—ten in all, radiating up and out from the mouth muscle.

So much for the direct results of smiling; now for the indirect results. The shortening and thickening process provides just the right antidote to the two major folds of flesh from nose to corners of mouth, by helping to lift them. In an earlier chapter I called these "the middle-aging lines." Then, too, all the light tensions of the cheek muscles (also, as we shall see later, the ear muscles) constitute a modifier or preventive of the jowls. If cheek and ear muscles could be sustained in tone, the jowls would never appear.

When the conscious mind finally gets the idea of controlling the cheek muscles and keeping them in a state of light tension all the time, we may forget about the smile; that was only good for a start. An artificial smile—awful, let's have none of it. But the expression resulting from continuous light tension of the cheek muscles will be just one of friendliness. Sometimes we are so accustomed to going around with drooping face muscles that even when there is a wonderful supply of friendliness in the heart it does not get expressed in activity of the face muscles. It always gets expressed somehow in the flesh—drooping or not drooping—the flesh itself may be insensitive, unaware; or, again, it may be sensitive, illumined. What we are trying to do here is to get back to expressing, in muscular activity, the natural friendliness that we feel.

The marvelous come-around of this thing is that the act of trying to express friendliness in the muscles, speaks to the friendliness that is in the heart all the time and encourages it to a fuller flowing. Try it and see.

I have the idea, too, that a face that goes around physically ready to smile is the foundation of the dazzling smile—no yard of slack to be reefed up before the smile can take off—just flash—there it is.

Before going on I would like to bring to your attention the customary misuse of the word *tension*. We use it as though it were *supertension*. But there are all kinds of tension: right, insufficient and excessive. Only excessive or supertensions produce ugly results. The mouth held in form by light tensions of the cheek muscles is an example of right tension. It is just like the light tensions required in good body posture: shoulder blades together and buttock muscles pulled in—light tensions there bring the lovely upright posture. Again take your mirror and we will see what we can do about cheeks and jowls.

The Santa Claus Exercise

Make a big smile, lips lightly closed; see if you can achieve the round, fat, smiling cheeks that belong traditionally to Santa Claus. Relax. Do this exercise like a drill several times, following with re-laxation.

Be sure not to pinch up the eyes when you do this. On the contrary, make simultaneous use of the skullcap to draw the eyes wide open, smooth out the forehead and lift the eyebrows.

Now start from the beginning and do the whole sequence of face lifting exercises to date:

1. Remind the skullcap to do its work:
   by means of the scalp lock move the cap,
   slowly backward and forward; relax. Do
   this a few times.
2. Then, *as the cap draws back*
   Let it lift and stretch wide the eyebrows—
   stretch them wide,
   stretch them wide;
   Let it open wide the eyes—
   open them up,
   open them up;
   Let it give the nose its upward lift—
   lift it up,
   lift it up.
   Along with the movements listed above now
   synchronize the action of smiling:
3. Make a pair of round, fat, smiling cheeks—
   lips lightly closed.
   And, by means of #1, 2 and 3 together—
   lift the whole face.
4. Relax.
5. Lift the whole works again in the same way.
6. Relax.
7. Now is the time for your sense of humor.

To do this exercise well requires practice; keep at it. Results do not come all at once, although I believe you will notice a difference right away. After a while as you get more and more expert you will discover that you are able to involve the lower part of the eyelid muscle, too. A good way is to brush up on a few functional winks *before* doing anything else. Once alerted, this area will, after a while, join in by itself.

An interesting point that you will begin to notice after you have worked with this drill for a while is the manner in which the action serves to sustain, or restore, the form of the upper lip. It is often the sag in the length-of-upper lip over the red-of-upper-lip, that conspires to change the red part into the thin line of "age." This does not mean that the older faces should try to sustain the mouth of youth (a biological impossibility); but that the mouth sculptured by over-relaxation and supertension is the mouth we do not need to have once we understand what the situation is. Nature, functioning through muscles rightly balanced and under the influence of a kindly spirit will bring changes to this feature, yes, but these changes will be lovely to behold.

And let us never forget that the smile and the laugh are the flying buttresses of the nose. If the nose does not have the support of the cheek muscles *and* the lift of the skullcap, then gravity takes over unimpeded and the nose weighs heavily all alone there in the middle of the face.

Summary of Exercises for the Mouth

For specific exercise, not posture at all, just occasional exercise to tone the mouth muscle we do well to whistle. And for lifting the mouth and firming the cheeks we do well to sustain the cheek muscles in a condition of light tension.

But the best workout for the cheek muscles comes as the Santa Claus exercise, for that contains within it all the elements of the other chapters to date; and latent within it the still undescribed and so important activity of the ear muscles.

The Mouth of Youth

Consider for a moment the face of a beautiful young woman, and focus your attention on the mouth. "If she is very young and as yet untouched by suffering, her mouth, regardless of its basic shape, will look soft and vulnerable";* but what are the years going to mean to

---

*From my friend and literary critic, Tish Thomas.

"are the years going to
mean to her — joy,
tenderness, humor";

"or perhaps disappointment,
bitterness and pain?"

(Top) A redrawing from a sketch by Edgar Allen Poe of his fiancee, Elmira Royster, who became the "lost Lenore" of his poems.

(Bottom) Drawing from a photo of Elmira at the time of the second courtship, twenty-three years later. She was then Mrs. Shelton and a widow.

The inside, intangible aspect of us is all the time at work building the outside tangible part.

CARDINAL GIBBONS: Courtesy Reverence For Life

her—joy, tenderness, humor; or perhaps disappointment, bitterness and pain?

The mouth dramatizes clearly that the inside, intangible aspect of us is always at work building the outside, tangible part. But what is so obvious at the mouth is equally true, of course, of the whole face, of every feature, not to mention every cell of the whole body. But the time has come now to draw another conclusion. This process of face and body being molded by the inner man we now see, can also work in reverse: what we undertake to do with the conscious mind about our faces, can and does turn out to be a source of influence upon the personality.

Now we are ready to proceed with a will to our last chapters—the first one is about the most unexpected place of the ear muscles in face lifting, and the second is about the throat—the poor old, beautiful, misunderstood, left-out-as-the-years-go-by throat. Lastly we will see what we can do about face and throat together by coming at them from the point of view of the body as a whole. At the end I shall tie up the whole works with a little bundle of what-to-dos that take no time at all to perform once you get the idea rolling. The fact is that if you never did one of the drills suggested in these pages, just finding out what the situation is, would constitute a big stride in the right direction.

Chapter 7

ENGINEERING WITH EAR MUSCLES

In all probability the ear muscles are a more important link in the overall strategy of face lifting than the general reader has any idea of at the start. But let me begin by describing the position of ears in the design of the face.

Since we are dealing with a feature that is around in the side plane of the head, we are confronted with problems of perspective. So when I ask you now to look at any well proportioned head, I shall suggest that you do so straight from the the front and on a level with your eyes. In this position you will notice that the ears are exactly in line with the nose—top for top and bottom for bottom they cover the same center third of the oval. This is the classical position of ears; but their story in the design of the face does not stop here.

The dynamics, life or movement of the face pattern owes a real debt to the position of the ears: for one thing it gives the pattern depth, three dimensionality. What is more, there in the outfield, these remarkable forms, along with the action lines of eyebrows and hair, serve to keep the eye of the observer roving over the whole face area—not just dropping like a piece of blotting paper on the stars of the show, eyes and mouth, but travelling around to take in the lesser players. Out there by the ear, the observing eye catches the strong, fine line of the facial contour, which, in turn, leads on down the cheeks to the chin; then up to the mouth, the eyes, and on around again. . . . As you watch television—and what a place to study faces!—watch for the ears, they are very interesting.

If the structure or shape of the ears offers a real problem the plastic surgeons can often help. There is some slight recourse in terms of the muscles, to help flatten the ears to the sides of the head; but ear muscles are powerless to change the shape of the ears. They serve another function altogether as you will soon see. However, it often happens that minor problems of ear shape can be alleviated by a deft arrangement of the hair.

These outside organs of the sense of hearing, when well made, as they usually are, have much beauty; they are a distinguished feature and are meant to be seen. Earrings, well chosen in relation to the proportions of face and throat, often contribute a rewarding accent to face design for dress occasions. And now to our main business with the ear muscles.

The Surprising Place of Ear Muscles in Face Lifting

There are three sturdy little muscles for each ear. In the anatomy books they suggest picture wire hanging the ears up on the side walls of the head.

What these small muscles do for us: All three, each in its own way, serves to *lift* the ears. The one closest to the face draws the ears up and forward; the strong one in the middle draws the ears straight up; the one toward the back draws the ears up and back. We use only the up and back ones for our face lifting. In those high and far off times when every rustling leaf and crackling branch might have meant the presence of a deadly enemy, we availed ourselves of these muscles. We drew the ear's bowl this way and that to catch the faintest vibration. The lessening of the need for such listening has been accompanied by an abandoning of the function of the ear muscles. However, the activity that we do not need, in these days, for listening, we can recapture and set to work in the interests of face lifting.

Though the human race no longer makes use of the active function of these muscles, they do not atrophy as do most long time abandoned muscles; they remain in perfect condition ready for work. It may be that their passive function of hanging up the ears has been enough to keep them going.

Ear muscles have two important connections in the world of the face muscles. Above, by means of the muscles of the temples, the ear muscles are laced solidly into the skullcap. And here follows a very important point: Ear muscles and skullcap work together as one action. Also, at the side of and below the ears, the ear tissues are blended in with the tissues of the cheeks. So it is that the muscles all the way up and down the sides of the face form a chain, and lifted scalp and ear muscles help to keep the flesh of the face fitted all snug and neat around the jawbone.

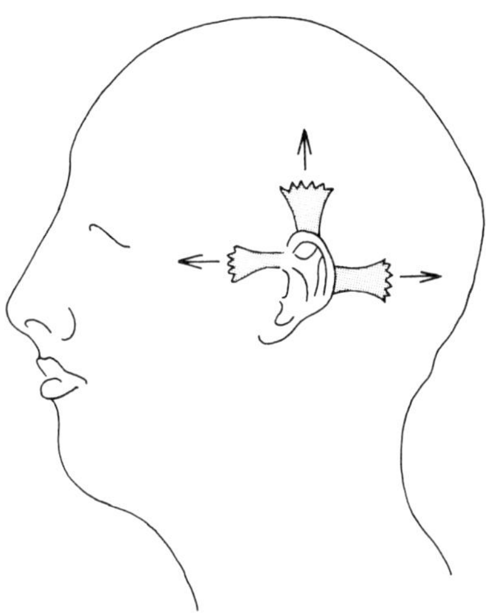

THE EAR MUSCLES

The Puller-Up
The Puller-Back
The Puller-Forward

The face lifter makes use of: the Puller-Up and the Puller-Back.

Pull them up and back the way a dog does when he is listening. Work them right along with the skull cap.

You already know well what the function of the skullcap is, to lean down over the crag of the frontal bone and draw the face straight up. And now we can add to this strategy the function of the ear muscles:

> Situated as they are half way up the face and to the side, ear muscles are in a position to provide the indispensable, supplementary pull, the up and back pull, so much needed for good face posture.

If you happen to be one of the people who is already able to activate, wiggle, his ears, you can make a little experiment for yourself. Watch in the hand mirror how the up and back pull of the ears lifts the loose flesh over the jawbone—the jowls—if it happens you have any. And then, also, notice the big crease from nose to corners of mouth that in an earlier chapter I called the *middle-aging lines* (though everybody has them, even a baby). Do you see how these folds draw back slightly, as in smiling—like the curtains of a miniature stage?

Reactivating the Ear Muscles

But if, at the moment, you find yourself unable to move your ears, place the tips of your fingers at the sides of your head where you know the ear muscles are situated. (Do not touch the ears themselves.) With this help draw the ears in an up-and-back position. As you perform this action, visualize the look of the muscles beneath the fingertips and think of them as achieving this action for themselves. Then tie in this up-and-back of the ears with the skullcap exercise as follows:

> Pick up the scalp lock with the right hand and, holding it close to the scalp, draw the scalp slowly back. Relax!

> Now synchronize the two movements—the back pull of the scalp with the up-and-back of the ear muscles; feel the whole muscular sheath that covers the dome of the head moving up and back—like a cap set jauntly at the back of the head. Relax.

> Repeat this double action with a change of hands for the other side. Do it over and over. Relax.

Setting these muscles to work is like learning to ride a bicycle: You just keep at it, and when you least expect it—away you go! Yes, resist gravity and all the down-dragging, in-pinching forces: keep the skullcap and ear muscles short and thick through using them.

Here I have some masculine testimony of an encouraging nature. This is from a man about forty-five, a friend who has been exceedingly helpful in the preparation of this manuscript. He writes: "I have been keeping my ears pinned back until it is almost a habit. I met some friends the other day who said: "How well you look, you look younger!"

And now before we move toward our conclusion let me affirm that functioning skullcap and ear muscles are the backbone of do-it-yourself face lifting. They are responsible for:

> the clear forehead,
> up eyebrows,
> sustained nose,
> well opened eyes,
> cheeks held back,
> jowls prevented.

When, in addition, we understand how to firm the cheeks with continuous light tension of the cheek muscles, we really do begin to have a grasp of the whole thing. (All of this is not a panacea. Perhaps it is just common sense brought to light with a rose pinned on it.) Contrariwise, when these three groups—cap, ears and cheek muscles—become elongated through lack of use they give the face its initial let-down, and the whole adds up to the face of the good old bloodhound.

Suppose that any one of us in middle life should go to the mirror and for the first time become aware of the flabby, sagging condition of the face. Each one remembering his physical training, is convinced that *flabby* is the opposite of *in-tone*. And in-tone has but one meaning for him at this moment—exercise! So, certain of the procedure's logic, he sets about making ghastly faces in order to put the flabby muscles to work.

But if we reflect upon the facts that have come out in these pages we realize why this would be a wrong step for the face lifter. Remember that tension and relaxation are a pair of opposites (as are also supertension and over-relaxation), but *in-tone* is the work of neither one side nor the other of this pair. And here follows a point of immense importance for us: *"In-tone" is made out of tension and relaxation, together, in perfect balance!* For the occasional muscle that would benefit from the face-making there are others for which this drastic treatment would be only intensification of activity far advanced in the wrong direction. Lines and creases would be deepened, and cords pulled further into prominence. We have found out that our face lifting technique is too subtle to be approached in this haphazard manner. We have first to learn what good posture is, and then practice it. Our researches to date have revealed that good posture for the face muscles is a twofold process, briefly: a relaxing-sustaining of the muscles of the face. The result of this muscular performance in terms of expression is one of cheerful serenity—a good base for all face expressiveness.

Chapter 8

THE THROAT

The throat muscles, though not muscles of expression at all, are still rich in expressive quality and must be given their share of attention. In terms of design the throat is to the face as stem to flower; the two together make up one visual unit. How many times do we see a face still sparkling with youth, and a throat already showing pronounced signs of age. This is hard to accept and, perhaps, not altogether necessary.

I once saw a young actress give an outstanding performance in the role of a blind girl. At the close of the last scene she was seated in a group of blind people facing the audience. The "tag" line had been spoken and in the silence that ensued, the girl quietly lifted her head. The movement was full of emotional impact for the audience: coming when it did, it became a symbol of the blind searching for the light.

To the uninitiated this action would have been nothing more than a lift of the head. But I would like to point out how much of the expressiveness came from the throat alone. There is a dynamic quality to the lines and masses of the neck; these forms are unmistakably moving from a lower to a higher plane. The correct timing of the full visual effect of the throat was capable of producing in those who saw it a moment not soon to be forgotten.

The Crucial Throat Muscles

Let us make no mistake about it, the muscles of the throat are not governed in the same direct sense as are the muscles of the face, by our thoughts and feelings; they are governed by the laws of the body as a whole and it is in this direction we must seek for our answers.

The Platysma Sheath, or Throat Upholstery Muscle

This sheath completely covers the front of the throat making all the various forms beneath appear smooth and beautiful—or so it does, ideally, for the women. With the men, the Adam's apple, or speech box, is still to be seen tucked neatly in between the cords of the *sterno,* or "bonnet strings" muscles. The platysma muscle comes in two sections that meet in the middle like a pair of closed curtains.

Attachments: The upper extremities of the platysma blend with those of the face all the way around at the jaw line. Some of these tissues make their way up as far as the grin muscles. Here we really can say that the throat is being lifted by the smile! The lower part of the sheath carries on down over the collar bones and blends with the chest tissues.

What the platysma sheath does for us: This muscle's function is, primarily, to help draw the head forward; (it also bares the fangs—but we will hasten over that vestigial fact). In our day this muscle is to be considered almost entirely in terms of its beauty potential.

As for the other neck muscles; at each side of the platysma we have the comely "bonnet strings" (see Michael Angelo's David, page 54) which turn the head right and left; and at the back, the strong trapezius that helps support the head and draws it backward. (page 99)

Now a few words on body alignment and body balance. A good place to begin an inspection of body alignment is at the base of the neck at the back. From this point the head is held aloft with unconscious ease and grace by means of the spinal column, the upper reaches of which are the seven cervical vertebrae. Below the base of the neck, the flat dorsal curve drops neatly down, with the curve of the lower back tucked well under it.

This lower curve, the lumbar curve of the spine, is held in position, at the front, by the lower part of the long abdominal muscle, and at the back by the buttock muscles. More on these muscles in the next chapter. (page 111)

In the rightly balanced body the neck muscles are occupied exclusively with their own particular function, that of supporting the head and rotating it in all directions.

But when the body is out of balance (muscular supertensions at one point or another often the immediate cause), then muscular compensations take place throughout the whole body—even the organs can be affected. Then the neck muscles are called upon to

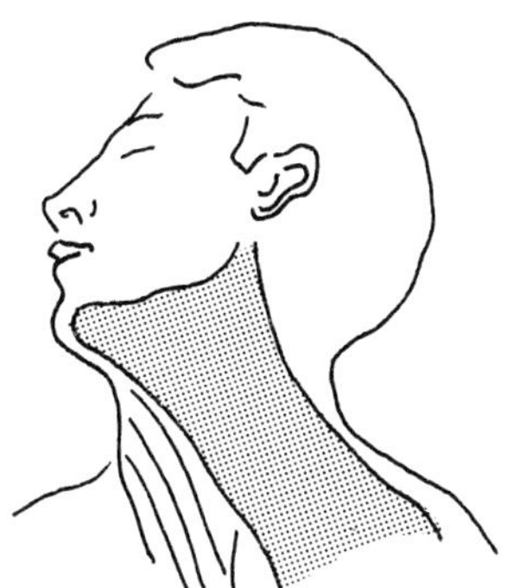

Platysma Sheath

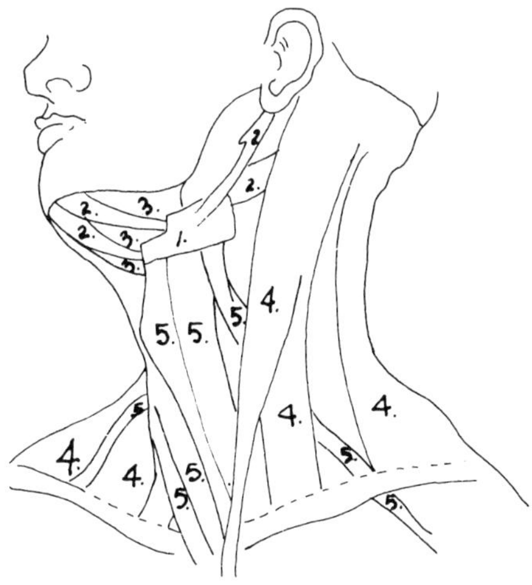

TONGUE MUSCLES

1. Hyoid Bone.
2. Muscles that contrive to thrust the tongue out.
3. Frog muscle that forms floor of mouth.
4. Great neck muscles that support and turn the head.
5. That draw the tongue down.

The best interests of the throat lie in good body posture, right exercise, balanced nutrition and all the other elements that make for health of the body as a whole.

help in the rigorous task of keeping the head in balance on top of the body. This unnatural strain shows up to a pronounced degree in the platysma sheath, pulling its delicate tissues out into the cords and folds we know so well and like so little. But when the spine and all its parts are in balance—then the throat upholstery muscle can be exclusively occupied with its own function and have time to spare for rest and recuperation.

So we are facing a conclusion of great importance to our theme, and one that, so far as I know, has not yet been sufficiently stressed from the cosmetic point of view; that the appearance of the throat as the years go by is profoundly influenced for good or ill by body posture.

A Study of the Double Chin or "Frog" Muscle

The *frog* is the second misery-making muscle on the front of the throat. (Not that all the trouble with throat appearance is related to muscle; fat of course plays a part.) The frog is a tongue muscle; you can locate it by pressing the backs of the fingers of one hand against the under chin area as you thrust the tongue out and back. Since it is tongue muscles that help provide the padding underneath the platysma we will see now if we can find out anything about them that might help us with the throat's appearance.

Singers can help us most about tongue muscles; they tell me that these muscles should be released from supertension. Finding out how they work will be a first step toward achieving this end.

The tongue is securely set down upon a little U-shaped bone that is situated at the front of the throat exactly at the junction of head and throat. You can find it, presenting its back to your fingers, if you feel around in that area. This little bone is called the *hyoid*.

The tongue muscles all have one end attached to the hyoid and the other to some strategically located bony prominence in the near, or not so near, neighborhood. By contraction, therefore, these muscles serve to manipulate the tongue, and by so doing, assist in moving the food about in the mouth, in swallowing, in speaking and singing.

One pair of these little muscles proceeds down the front of the throat, their function being to draw the tongue back and down. The pair that pulls opposite to these, through an engineering feat that takes your breath away, manages to hurl the tongue out of the mouth in a manner, someone suggests, akin to the functioning of

the medieval battering ram: you pull on the ropes and, wham, out
goes the big log to knock down the walls.*

Out of the whole fascinating bunch of tongue muscles it is the
following pair of tongue retractors that turns out to be the most help
with the double chin. Like the ones mentioned just above, this pair
proceeds down the length of the throat; but they do not stop there
as the others do, they continue right on down, cutting cross-lots
through the chest and fasten at last, one on each side to the tops of
the shoulder blades! Here we find a second tie-up between the
appearance of the throat and body posture. When we take up
good body posture—head high, chin in, shoulder blades together,
buttock muscles drawn tight together and under—we help not only
to keep the platysma sheath all smooth and lovely, but also to hold a
checkrein on the double chin.

So, good body posture turns out to be an essential, indirect, approach to the improvement of the throat's appearance. I have coming up now some important what-to-dos relating, directly, to this
crucial area. The next in line I call—

First Aid for Throat Muscles

Lie on your bed flat on your back. Put a pillow under your
shoulders—under the cage of the ribs actually, so that your head
droops backward over the pillows. Place the fingers of one hand
lightly over the throat so you can feel the muscles there getting a
sound workout as you:

> draw the head up;
> lower it slowly back;
> relax.
> Repeat:
> draw the head up;
> lower it slowly back;
> relax;
> up; down;
> up; down.

---

*Practicing tongue-out and then back-down in the throat, would seem to provide
a logical way of helping keep good tone in the throat muscles.

Do this five times slowly and rhythmically with deep breathing. Inhale deeply just before the up pull and exhale on the release. Be faithful, without overdoing, and you will greatly help to firm the muscles of your throat and keep them in tone. This exercise would clearly be at its most effective if carried on through a lifetime as part of a program of exercises for the body as a whole. We must not expect to undo lifelong wrong muscular habits in a few days. The old proverb—slow and steady does it. And now follows:

The Cold Water Cure

The woman who told me about this "passive" technique had done exceedingly well with it herself and that is why I pass it on to you.

Fill a basin with cold water. Then take a small turkish towel, about fifteen inches by twenty-five. By folding it across the length, which gives roughly a square, and then three times more, you will find yourself with a thick bandage long enough to reach from ear to ear under the chin.

Starting at the base of the throat and working right on up to the top of the forehead, sop on the cold water for five minutes. Keep the water as cold as possible. When you have finished, and glance up into the mirror you will see how well the tissues of the throat and face are being fed with the blood food prepared for them in the body laboratory.

Results: The Cold Water Cure causes various good things for throat and face to happen simultaneously. First, these tissues, full of nerve ends, are fed and revitalized. Then, too, the blood that has been stirred up in this small area will, in the ordinary course of events, be pumped all over the body. So what we do for the face and throat serves also to refresh the whole of us.

Let me tell you now about a most wonderful friend, a Scottish lady, a retired head nurse, who did better with her throat than any other human being I have ever known. She had a youthful figure and a pretty throat at seventy. Her posture was excellent. When I questioned her as to what she had been doing to keep her throat in such good shape, she told me that she had always been interested in matters of physical culture, including nutrition, and that through the years she had steadily practiced the head rolling exercise. She proceeded to roll her head around. So let me now line up head rolling as a regular drill.

The Head-Rolling Exercise

For this drill you should sit straight, head erect. Whenever possible, do it in the sun.

1. *The vertical roll:* Now roll the head all the way up for a glimpse of the ceiling, then all the way down, chin on chest, for a glimpse of the floor; up, down; up, down. Return to the starting position.

2. *The sidewise roll:* Next, roll the head across the tops of the shoulders. This is a sidewise movement exclusively. Do not drop the head, even the slightest, either forward or back. We could think of this one as the pendulum of the grandfather's clock swinging: tick, tock, tick, tock. Return to starting position.

3. *Round and round:* Lastly, drop the head forward and roll it all the way around to the right, and then all the way around to the left, five times each way or alternating.

This drill accomplishes many fine things. By limbering the ball and socket joint where head joins spine it helps break up the bad posture habits relative to the carriage of the head, and by so doing relieves the nervous tensions centered there; also it provides a balanced workout for the throat muscles. And thrown in for extra on sunny days, it brings the healing sunlight to the eyes, the sinuses, the throat, the chest and the whole person.

The Throat as a Way-station
Stop, Look, Listen!

The period of middle age is the great warning time, the great preventive medicine time, the time to take a serious look at the effect upon the face, throat and body of the way we are living our lives. Men and women would do well to take their concern over a face and throat that are beginning to show signs of "age" as an arrow pointing to the need for toning up the condition of the body as a whole. Are you having enough exercise in the open air? Are you breathing deeply and getting enough oxygen into the system—oxygen, man's very lifeline! Are you getting the right foods for health?* Are you having enough rest and recreation? First of all,

---

*Davis, *Let's Get Well, op. cit.,* Chapter 32, *Two Unvarying Rules;* and Chapter 33, *Planning Your Nutrition Program.*

to go to the very root of the whole business—do you take time to refresh yourself mentally and spiritually, for man is *one*. All of which carries us on to the next chapter.

Before we move on, however, I have a few brief suggestions about *costume in relation to the "aging" throat* which may prove useful while we are practicing the more important things.

After what has been said in an earlier chapter about necklines nothing need be added here except a warning. We must make doubly sure that we do not, inadvertantly, bring special emphasis to this area with a harsh neckline, or one too low, or one that accentuates rather than counterbalances off-standard facial proportions.

When it comes to color: Color is for everyone, surely, perhaps especially for older people. Color and white hair can be pretty scrumptious. But, close about an "aging" throat—always—soft shades, soft textures, and unobtrusive lines will be found the most becoming. Keep this area well designed and quiet, and let the face, with its uniquely exciting few square inches of living pattern, challenge the attention of the observer.

Chapter 9

THE BODY

We move now from the world of the face and throat to the world of the body.

The Way We Wear Our Flesh

Flesh is an instrument of expression in the same sense as are eyes, hands, and the total body. The entire quality of a person's character is there in the flesh, to be seen with the eyes and touched with the finger.

Today we are more likely to recognize life's rare and wonderful people by a quality of the flesh than whether or not that flesh is lined or drooping. And the reason is this: some people outgrow their faces; in these faces we see the marks of what their wearers have been through, the "aging" characteristics set there primarily by the overtensions of anxiety. But right along with the telltale signs of the human struggle we see with delight a certain exquisiteness of the flesh. If one of the values that we find so attractive in the younger faces is the firm "up" of the flesh, the unique aspect that we find in the occasional older faces is this lovely quality of the flesh itself. There are some few people who have both of these characteristics together. Through such people as these we find that the double beauty—the "upness" and the sensitivity both together, is a possibility for all of us.

But now let us look into this subject more in detail. Flesh, except for the fat, is composed chiefly of muscles and skin. Since we are relatively well acquainted with the muscles, let us now turn to the skin. (1) Skin is the elastic outer covering of the body. (2) It is the organ of the sense of touch, no less. (3) Where it covers the face it may be described as a feature, and a very important one.

Among other things, the skin is also the bearer of the pigmentation. A human being's color, unlike that of the carrot, which is yellow-orange all the way through, or the tomato, which is red all the way through, is the exclusive property of the skin. How stunning, how exquisite are the colors!

Then the skin has two other important functions: *The skin breathes:* If we should cover ourselves from head to foot with a paint that completely stopped up the pores, in a few minutes we would be dead. In the early days of show business just such a tragedy occurred. (We may rest assured that body makeup today, prepared as it is especially for the human skin carries no such threat.)

The skin is an instrument for eliminating body poisons. I shall never forget the summer I worked on a farm—sweating it out that hot July in the hayfields. When I got back to town I felt as though I were walking on air and people said to me: "What have you been doing, you look like a new person!"

A Few Fundamentals of Skin Care

Cleanliness

In view of the skin's functions—breathing and eliminating—it follows that cleanliness is the first step in skin care. Since individual skins differ so widely as to oiliness and dryness, I shall make no recommendations here for the cleansing. Each person knows what works best for him. However, one point is worth noting. The soaping that the men put their faces through in shaving seems to give them an edge over the women. I believe all of us who use soap delight in the wonderful cleansing that may be had in this way. However, it does leave the skin in need of further attention, and here is the place for our next consideration.

Lubrication

Lanolin is a product of sheep's wool and contains an oil more closely resembling the skin's natural oil than any other product. To call lanolin a skin food is a misnomer. Skin works, almost exclusively, in the other direction—it *excretes*. Lanolin is a lubricant, and for this purpose it is still, for many, unexcelled.

Application: When rightly prepared for toilet purposes lanolin is soft and easy to put on; a small amount of it worked into the creases and a very small amount over the whole face and throat will be found helpful and comforting after a cleansing with soap.

Warning: Because lanolin is so relaxing, it should not be used as a remedy for sagging tissues. We know what sagging tissues are calling for—just the reverse, a tightening up. We must confine the use of lanolin to the minimum requirements of relief from dryness. Then too, in this connection there is a special warning about the nose: this feature is so generously supplied with grease glands, that, in most cases, it is better off without additional lubrication. What the nose needs, often, is a cleansing of the pores with a soapy lather.

Care of the scalp: Now and then, as we feel the need, we would do well to moisten the fingertips with a minute amount of lanolin (more or less according to dryness) and for a few moments give the scalp a brisk massage. This will set the circulation going and aerate the roots of the hair. In my own experience the hair responds to such attention almost at once. It stands up better and looks more alive. Beyond this ministration the means to healthy, youthful, vigorous skin and hair would seem to lie, very largely, in right foods for the body.

*Nutrition—The Wonder-working Foods**

Let me ornament these pages with a few of them:

The Fresh Fruits:
These are the cleansers and youth sustainers.

The Fresh Vegetables:
These are the protectors, so valuable for their vitamins and also their bulk. We do well to eat as many of them as possible raw, in or out of salads, or as juice—at least one raw vegetable with every meal. If we keep the colors coming in great variety the mineral balance will take care of itself.

The Proteins:
These are the body builders, meat, milk, fish, eggs, soy beans,

---

*People on a medically restricted diet will, of course, follow their doctor's instructions.

cheese, nuts. With the proteins, too, we do best to keep them coming in great variety.

The Starches and Sugars:
When these are of the right kind and eaten in the right amounts, they are energy makers. Starting with the sugars, the natural ones built right into the fruits and vegetables are the best. Then come the unrefined sugars that still carry their own vitamins. Honey is a wonderful, healthful sweet. As for the starches, the best flours are the ones that contain the whole grains and corn—the staff of life.

Surely there is no one to deny that right foods eaten in the right amounts, and the right way, do a big share to bring the glowing colors of health to skin and hair and sparkle to the eyes. Such wonders have been accomplished for these features—skin, hair and eyes, by means of right nutrition that anyone having trouble with any one of these would do well to make a thoughtful reading of Adelle Davis' *Let's Get Well.** Besides having a vast fund of information, all medically authenticated, Adelle had a sense of humor and that helps anywhere.

In our day the medical advance guard is in the midst of discovering that rightly balanced foods have a dramatic share in producing and sustaining healthy mental and emotional states. Great developments lie ahead on this frontier. Since good nutrition plays a vital role not only in health of body but also in health of mind and heart, it plays a double role in the production of beauty.

When it comes to the subject of *overweight*, we have to recognize that, in many instances, the cause of this condition lies in starvation for the right foods. Starvation for right foods, alone, can produce abnormal cravings for sweets, starches, alcohol, cigarettes, drugs and other means through which the individual attempts—with or without realizing it, and always without success—to compensate for the chemical lacks of his body.

But if the weight watcher transfers his attention from the painful business of cutting down on foods, to having plenty of the right foods, he will soon find the surplus pounds as well as, oftentimes more serious maladies, disappearing.

**Op. Cit.,* Chapter 12, *Skin Problems are More Than Skin Deep.*

In periods of reducing we must not overlook the fact that one of the characteristics of the healthy skin is its *elasticity;* and this, obviously, means a capacity not only to stretch out, but to pull back again. Our basic approaches to the muscles and nutrition, together, help us mold and remold the face tissues snugly around the face bones where they belong. We must not be discouraged—finding a right direction, in itself, leads to beauty—making peace and joy.

The Way We Carry Our Bones

I want now to develop further the so important matter of body posture. This I do, not only because all poor posture shows up in face and throat as strain, but also in order that face and throat together may find themselves set like a work of art in a body framework that can be counted upon to enhance them.

Probably you have watched a child make a representation of the human figure in clay. He makes two balls, one larger than the other and pops them together for head and torso. In similar manner he attaches the arms and legs. These first interpretations of the human figure have no neck; the head is dumped down upon its base like the head of a snowman.

Much of the body posture we see about us looks as if it were founded on this childish understanding of body structure; but an entirely different posture results when we not only understand our anatomy but take a lively interest in its functional use. I have spoken of gravity as one of the significant pullers-down of the face; and so it is, also, for the body. We should not forget, however, that gravity is the setter-up of the whole scheme of life on this planet. Body balance exists in relation to the pull of gravity.

There is a fresh angle on the old story of good posture that ties in especially well with our approach to the face muscles—a technique that, widely as it has been known and celebrated in Europe, deserves to be far better known in America and all over the world. This technique is the *System of Functional Exercises* created by Mrs. Bess M. Mensendieck, M.D.* I propose now to discuss in my own way two

---

Look Better, Feel Better by Bess M. Mesendieck, M.D. New York: Harper and Row, Copyright 1954. Credit should also be given to two of Mrs. Mensendieck's finest teachers, Miss Paula Pogani and Mrs. Constance Taylor with whom I studied over a period of many years.

of Dr. Mensendieck's exercises that I believe, most succinctly, go to the heart of the erect posture.

Two Important Pairs of Posture Muscles

1. The Great Buttock Muscles

These muscles form the cushion of the body's seat. When they are ignored throughout the years they grow flabby, thereby contributing to the general appearance of droop in the aging figure. But, far worse, lifeless buttock muscles mean that we have lost our grip on good posture—with all the tragic implications of that, including a less than tolerable appearance of the throat. But when buttock muscles stay on the job, they are round and firm and contribute greatly to the body's youthful stance.

The Disciplined Pelvis

If you look carefully at the torso part of the skeleton, you will be reminded that it is made up of two bone groups (shelter boxes for organs)—the rib cage above, and the pelvic girdle below. These two bone centers are held together by the flexible spine and, also, of course, by the still more flexible muscles. *Good posture requires that the pelvis be held well in and under the cage of the ribs.* Two muscles, working in close cooperation, are responsible for sustaining the two bone groups in proper relationship to each other; they are, at the front, the long abdominal muscle—lower section—and, at the back, the buttock muscles. Please identify these two muscles on the diagrams. Then we are ready for action:

The Pelvic Roll

Sit straight.
Place both feet on the floor in front of you.
Then contract the abdominal muscle at its base, and the buttock muscles, both at the same time.
By so doing you draw the pelvis into position—in and under the cage of the ribs.
Relax, breathe deeply.

As you do this you will make a discovery; that the movement does not seem to be a concluding one—it tempts you to proceed. The back comes up, the shoulder blades close together, the head rolls up into position (chin in) and the posture is left tall and beautiful.

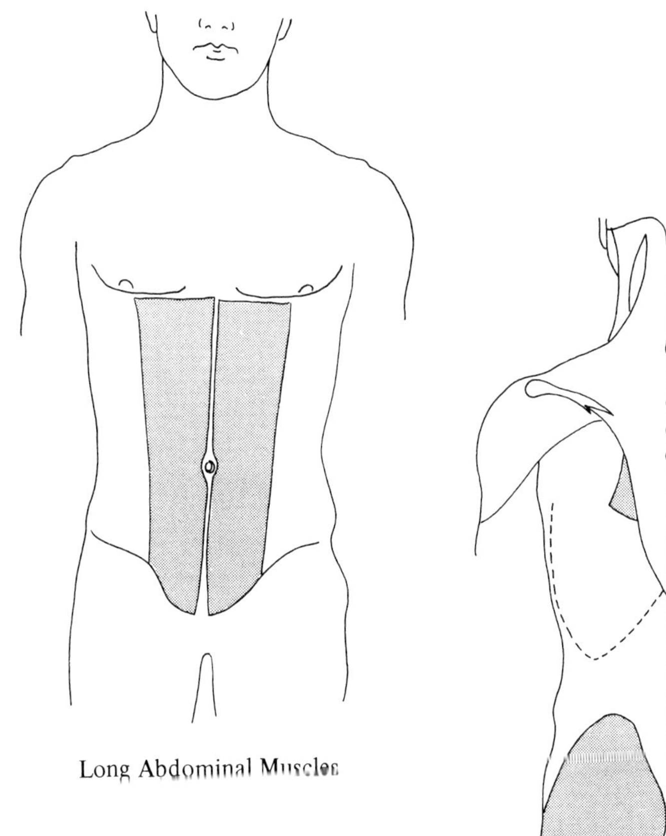

Long Abdominal Muscles

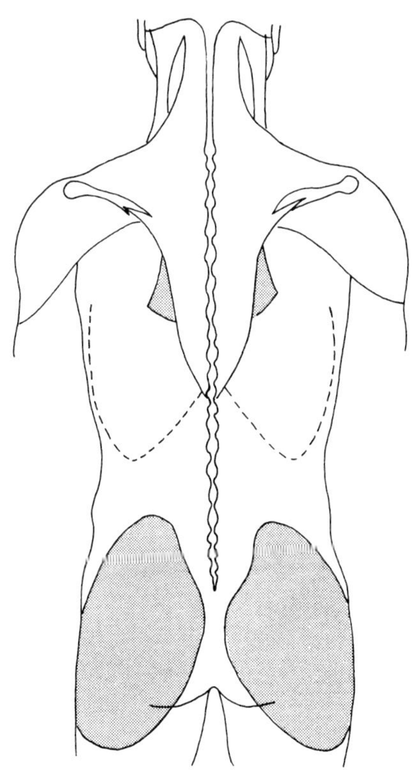

Buttocks

A fifty-five-year-old friend of mine experienced an acute attack of osteoarthritis centered in the hips; it pulled her usually good posture so completely out of line that she waddled like a duck. But this pevic roll exercise that she had practiced through the years did not fail her. Along with other exercises prescribed by her doctor, she kept at it and at long last regained a normal posture.

2. *The Rhomboids or Shoulder-Blades-Together Muscles*

The art of the erect shoulders lies in understanding that this desirable part of good posture does not come as a result of attempting to *push* the shoulders back. It comes from *drawing the shoulder blades together* by means of a small pair of muscles designed for that specific purpose—the *rhomboids*. Study the diagram and see how these little muscles join the shoulder blades to the spine. They are a pair—one on each side. (p. 113)

The Shoulder Thrust

1. Place both feet together on the floor in front of you; sit straight and keep the back erect throughout the exercise.
2. Raise the arms straight to the sides, palms down.
3. Then extend the arms as far as you can to left and right; with the fingertips reach for the walls on either side. This action thrusts the upper arm bones out from the shoulder sockets.
4. Here now comes the crux of the matter: keeping the arms in the exact side plane (palms still down) *draw the shoulder blades tight together to the spine.*
This action returns the arm bones into their sockets. If you work very hard you close the shoulder blades tight together.
5. Repeat slowly several times: Thrust out, draw in; out, in; out, in—tight, tight, tight!
Relax. Deep breath.

We remember from our surgeon that unused back muscles elongate and serve only to hold the bent over posture. If your shoulders have already begun to develop a stoop you will have to go about this exercise very gradually—just a little effort at a time in the right

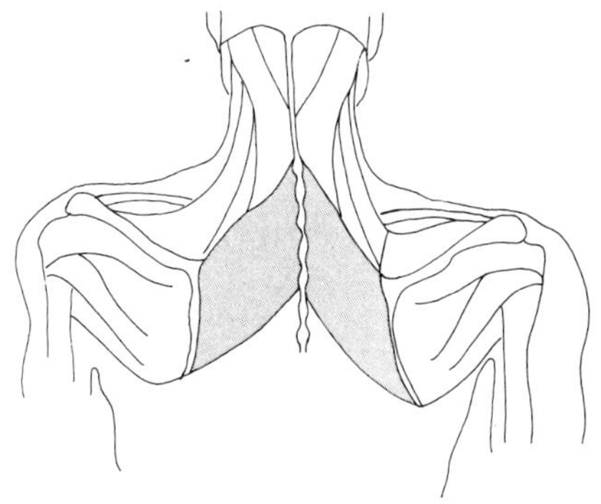

Rhomboids

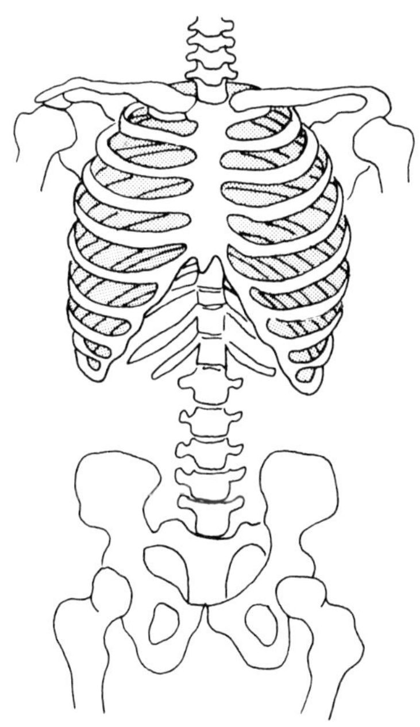

Bones of Trunk
(Front View)

direction. But if you make use of these little rhomboids, all the way along, the stoop will never come; and the practice will yield challenging side effects. It will produce a brave carriage of the head, improve the appearance of the neck and shoulder line and—by releasing the platysma sheath from supertension—it will help to maintain a smooth and lovely throat.

Since anxiety over the appearance of the throat comes along with the first miseries of adjustment to "aging"—may even be the factor that triggers these anxieties—you will find that the understanding which suggests: "I must look to my posture," has great significance not only for face and throat but as a checkrein on old age thinking. It converts the anxieties into constructive action. We know by now that it is *not primarily time* that does these unkind, aging, things to us so much as it is *what we think, what we feel, what we do.* If we have come to grips with this concept as a working reality then, surely, we will have "old age" in the palm of our hands.

Chapter 10

SUMMARY OF FACE LIFTING TECHNIQUES

1. *The Body Framework*

Take up good posture:
   buttocks drawn in and under;
   back rolled up;
   shoulder blades drawn tight together;
   head aloft;
   chin in;
   tall and stunning.

2. *The Warming Up*
Stand erect but at ease;
Take a deep breath;
Relax.
Do a few of the head rolling sequence: up and down, side to
   side, round and round.
Then do your *functional winks*. Do not squint. Do twenty-five
   of them, or more, as you wish.

3. *The Heart of the Matter*
Sustain good posture, but at ease, throughout.
Then do twenty-five *sanpans.*
A sanpan is the following activities done simultaneously.
   Scalp back;
   eyebrows high;
   ears pinned back;
   lips lightly closed, keep them closed.
Then tie all together with; a big smile. Relax. When you
have become proficient you can perform a sanpan in one
moment.
Do twenty-five more sanpans.

For those who have followed each chapter this synthesis will be easily understood—it is, by and large, the muscular activity of the animated, youthful, smiling face—the face of joy—consciously utilized as an exercise. A face rigged with live, functioning muscles will never blow down. The muscles are to the face what halyards are to the sails—heave away, sailors, and up she goes! Everyone loves to look upon a fresh morning face under full sail.

And now before we move on from our study of the technical effects of the happiness and joy group upon the human countenance I have one more important, very practical consideration to share with you:

When the joy muscles are active, the worry muscles are getting their normalizing back-pull, their stretch back, their beauty treatment. Obviously when the joy group performs its action often enough, the worry muscles are prevented from developing into posture creases. And let us not forget that what takes place in the face has implications, also, for the entire body.

How Good Habits Come Easily

Since one of the inviting aspects of our face lifting technique is that we have to set aside so little time for it, let us now reduce that *little* to a minimum, by proper planning.

The First Stage

In the beginning it does make sense to take the hand mirror and give a few minutes to observation and experiment.

The Second Stage

Then, knowing what it is we are after, we proceed to make the doing of it a continuous performance. The best times to utilize for this period are those when, though we have to be present, we can sit back and let someone else do all the work. Such times are those of riding in trains, buses, planes; listening to radio, watching movies or television. We can train ourselves to associate buttock-muscles-hitting-the-seat with face-muscles-lifting-up.

After a while you will find that the action of face lifting with its good posture is beginning to spread out by itself all through the day. You can feel it becoming a part of you. At this time the hours of driving a car provide a wonderful opportunity. A colored string tied to the wheel will be a good reminder until the association with driving becomes complete. No one will find himself any less attentive because he rides along holding his face up. On the contrary, holding up the face becomes a condition of alert.

No face-making that will attract attention need be involved. Whenever you think of it, not as posture at all now but definitely as exercise, practice that little Mona Lisa smile; acquaint yourself with its light tensions. Work on the skullcap via the eyebrow and open the eyes wide.

Very slowly you will find that you are developing the ability to draw the face muscles out from the plane of the involuntary into the plane of consciousness where you can order them around. In response to a flick of the mind you will be able to slip the face like a glove up from the throat, over the chin, up over the face bones, on over the dome of the head, to button it down, at last, at the back of the skull. We have seen that this is not a metaphor, it is a reality —the face, scalp and throat are one sheath of interplaying muscle, from the collar bones at the front to the occipital bumps on the back of the skull; we must think of it as pulling all the way up, and over, and back, and coming to rest there. We need to counteract all the down-dragging, in-pinching forces of life and keep the face ready to smile. Even when we have worked out all the problems of the body, mind, and heart, which will be quite an achievement, we will still have gravity to contend with. And the watch-word turns out to be: *Persevere!*

In closing now I would like to tell you a story. This episode comes from the same beauty shop I was telling you of in our opening pages. It is not about the woman with the redgold hair, but quite another—

The Visit of Mrs. H.

When Mrs. H. entered the studio I was busy preparing the place for the next client, but out of the corner of my eye I sensed that she was not the type who ordinarily finds her way into a deluxe beauty salon. And so it turned out. She was filling the appointment of a friend. The maid took her wraps and she sat down at the dressing table for the preliminary analysis. Together we looked into the mirror.

The lights on the dressing table fell full on the face before us. The beauty of it left me momentarily speechless. Surely this was the face of old age, there could be no mistaking that, the special wonder of it for me lay in the quality of the flesh, a youthful flower-like radiance that was indescribable.

It was clear enough, however, that we were not looking at the same thing. She made a little wry expression, and with a gesture of the hand as if to wipe away the years, said apologetically: "I am seventy-four."

In some ways this was apparent. She had many aging face characteristics. I explained to her as best I could in a few minutes, the principle of lifting faces by voluntary control of the face muscles. I also pointed out that the pattern for good face posture for these muscles was suggested to us by their activity in the expression of joy. She was fascinated. The mental turnover in her eyes was absorbing to see. The face in the glass that only a few moments ago had been telling her how old she was, had all of a sudden become an organic, growing thing to her, as full of possibilities as clay in the hands of a sculptor. Leaning toward the glass and trying to lift her face with a pretend smile she said: "I guess I do go around looking awfully serious." She received the complete treatment and obviously enjoyed every minute of it.

I tell you this episode just now because it reminds us of the so important subject—the talking flesh—the whole quality of a person's character is there in the flesh to be seen with the eyes and touched with the finger. But I have told it also for another reason: A few days after Mrs. H.'s visit to the studio I went to Grand Central Station to meet a friend from an incoming train. Searching the faces of the passengers as they came down the ramp, I recognized this woman at once, and she me. We caught one another by the hands like old friends: "What have you been doing?" I enquired, "You look so gay and wonderful!"

"Oh," she answered with a smile I shall not forget, "I've been holding my face up; I've been telling my family in Boston about it, how you have to start holding up your face!"

MONA LISA (detail) by Leonardo da Vinci, Louvre Museum, Paris. Photo by Alinari, Florence, Italy. Permission granted. With appreciation · of Fogg Museum, Cambridge, Mass., for their kind assistance.

READER'S NOTES

READER'S NOTES

READER'S NOTES

READER'S NOTES

READER'S NOTES